T0407743

ENVIRONMENTAL SCIENCE, ENGINEERING AND TECHNOLOGY

THE PEAT SWAMP

PRODUCTIVITY, TRAFFICABILITY AND MECHANIZATION

ENVIRONMENTAL SCIENCE, ENGINEERING AND TECHNOLOGY

Additional books in this series can be found on Nova's website
under the Series tab.

Additional E-books in this series can be found on Nova's website
under the E-book tab.

ENVIRONMENTAL RESEARCH ADVANCES

Additional books in this series can be found on Nova's website
under the Series tab.

Additional E-books in this series can be found on Nova's website
under the E-book tab.

ENVIRONMENTAL SCIENCE, ENGINEERING AND TECHNOLOGY

THE PEAT SWAMP

PRODUCTIVITY, TRAFFICABILITY AND MECHANIZATION

**ATAUR RAHMAN
RAFIA AFROZ
AND
AZMI YAHYA**

Nova Science Publishers, Inc.
New York

Copyright © 2012 by Nova Science Publishers, Inc.

All rights reserved. No part of this book may be reproduced, stored in a retrieval system or transmitted in any form or by any means: electronic, electrostatic, magnetic, tape, mechanical photocopying, recording or otherwise without the written permission of the Publisher.

For permission to use material from this book please contact us:
Telephone 631-231-7269; Fax 631-231-8175
Web Site: http://www.novapublishers.com

NOTICE TO THE READER

The Publisher has taken reasonable care in the preparation of this book, but makes no expressed or implied warranty of any kind and assumes no responsibility for any errors or omissions. No liability is assumed for incidental or consequential damages in connection with or arising out of information contained in this book. The Publisher shall not be liable for any special, consequential, or exemplary damages resulting, in whole or in part, from the readers' use of, or reliance upon, this material. Any parts of this book based on government reports are so indicated and copyright is claimed for those parts to the extent applicable to compilations of such works.

Independent verification should be sought for any data, advice or recommendations contained in this book. In addition, no responsibility is assumed by the publisher for any injury and/or damage to persons or property arising from any methods, products, instructions, ideas or otherwise contained in this publication.

This publication is designed to provide accurate and authoritative information with regard to the subject matter covered herein. It is sold with the clear understanding that the Publisher is not engaged in rendering legal or any other professional services. If legal or any other expert assistance is required, the services of a competent person should be sought. FROM A DECLARATION OF PARTICIPANTS JOINTLY ADOPTED BY A COMMITTEE OF THE AMERICAN BAR ASSOCIATION AND A COMMITTEE OF PUBLISHERS.

Additional color graphics may be available in the e-book version of this book.

Library of Congress Cataloging-in-Publication Data

The peat swamp : productivity, trafficability, and mechanization / editors, Ataur Rahman, Rafia Afroz, Azmi Yahya.
    p. cm.
 Includes bibliographical references and index.
 ISBN 978-1-61942-670-2 (hardcover)
 1. Peat machinery. 2. Peatlands. I. Rahman, Ataur, 1967- II. Afroz, Rafia, 1972- III. Yahya, Azmi, 1960-
 TN841.P43 2011
 577.68--dc23

2011052249

Published by Nova Science Publishers, Inc. + *New York*

CONTENTS

Preface		**vii**
Part 1:	**Introduction to Peat Swamp:** **Introduction, Productivity and Economic Values**	**1**
Chapter 1	Introduction: The Peat Swamp	**3**
Chapter 2	Productivity and Economic Value of Peat	**15**
Chapter 3	Economic Value of Peat Land	**19**
Part 2:	**Traficability (Mechanical Properties and Chemical and Biological Characteristice)**	
Chapter 4	Traficability of Peat	**35**
Part 3:	**Mechanization: Wheeled Vehicle, Tracked Vehicle and Intelligent Air-Cushion Tracked Vehicle**	**63**
Chapter 5	Mechanization of Peat Swamp	**65**
Chapter 6	Wheeded Vehicle for Highland Peat	**67**
Chapter 7	Tracked Vehicle for Moderate Peat Terrain	**89**
Chapter 8	Intelligent Tracked Vehicle for Worst Peat Swamp	**117**
Chapter 9	Semi-Wheeled Tracked Vehicle: "Highland and Moderate Peat"	**169**
Chapter 10	Relevant Peat Vehicle in Malaysia	**175**
Index		**179**

PREFACE

This book has been written with emphasis on the fundamental engineering principles underlying the peat characteristics, importance of peat, environmental effects of peat, determination of the mechanical properties of peat terrain, mechanization of peat based on the vehicle for highland, moderate peatland and peat swamp. The mechanization of peat has been discussed in regards to the development of peat vehicle and their performance. This book intends to introduce senior undergraduate and postgraduate students to the study of the peat terrain and theory of peat vehicle mobility. An intelligent control system is presented in this book with air-cushion tracked vehicle which provides a valuable resource on control theory of air-cushion vehicle in swamp terrain for all automotive students. (Imprint: Nova)

PART 1:
INTRODUCTION TO PEAT SWAMP: INTRODUCTION, PRODUCTIVITY AND ECONOMIC VALUES

Chapter 1

INTRODUCTION: THE PEAT SWAMP

1. THE PEAT SWAMP

Peat swamp is waterlogged land which is growing with dead leaves and plant material up to 20 metres thick. They comprise an ancient and unique ecosystem characterized by water-logging, with low nutrients and dissolved oxygen levels in acidic water regimes. Their continued survival depends on a naturally high water level that prevents the soil from drying out to expose combustible peat matter. This harsh waterlogged environment has led to the evolution of many species of flora uniquely adapted to these conditions. Peat forms when plant material, usually in marshy areas, is inhibited from decaying fully by the acidic conditions and an absence of microbial activity. For example, peat formation can occur along the inland edge of mangroves where fine sediments and organic material become trapped in the mangrove roots. Peat is mostly soil about 65% organic matter which is composed largely of vegetation including trees, grasses, mosses, fungi and various organic remains including those of insects and animals. Peat formation occurs when the rate of accumulation of organic material exceeds the rate of decomposition. The build-up of layers of peat and degree of decomposition depend principally on the local composition of the peat and the degree of water logging. The peat acts as a natural sponge, retaining moisture at times of low rainfall but, because it is normally waterlogged already, with a very limited capacity to absorb additional heavy rainfall during periods such as a tropical monsoon.

Peat swamps are an important component of the world's wetlands due to its dynamic link between land and water, a transition zone where the flow of water, the cycling of nutrients and the energy of the sun combine to produce a unique ecosystem of hydrology, soils and vegetation. These swamps provide a variety of goods and services, both directly and indirectly, in the form of forestry and fisheries products, energy, flood mitigation, water supply and groundwater recharge.

Figure 1.1 shows the typical swamp peat land. Peat swamp forests develop on these sites where dead vegetation has become waterlogged and is accumulating as peat. Water in peat swamps is generally high in humic substances that give a typically dark brown to black colour to the water. These conditions influence the types of vegetation that thrive in the covering forests and that, in turn, contribute to the character of the swamps. Peat swamps are characterized by diverse features that relate to the nature of the water supply, such as flooding by surface or groundwater, or solely from rainfall; type of landscape in which the peat swamp occurs, such as shallow depressions close to rivers; type of landscape that the swamp creates, such as accumulation of peat above groundwater level so that vegetation, often with prominent aerial roots, becomes wholly dependent on rainfall. Peat water is dark-brown to a murky black, and is acidic of pH-value in the range of 3 to 4. Peat accumulates in domes with a depth of 10 to 12 metres and flows from watersheds into the main rivers.

Peat swamp forests (PSF) have a specific atmosphere and many different animal sounds are heard. Temperatures within the forests are moderate and under closed canopies seldom exceed 28°C. There is a noticeable wind circulation in the afternoon. Soil and water have a constant temperature of approx. 23° - 24°C. Tree types and fish species have adapted to the acidic water. Special roots protrude out of the water to absorb oxygen (Page, and Rieley (1998) and Reiley et al., 1996). Figure 1.2(b) shows the swamp peat which is deforesting for agricultural land.

Figure 1.1. Peat Swamp Land.

(a)

Figure 1.2. (Continued).

(b)

Figure 1.2. Moderate peat swamp (a) clearing the vegetation (b) drainage for the agricultural purpose.

1.1. Importance of Peat Swamp

Peat swamp forests are unique habitats for fauna and flora, commonly with a high proportion of endemic species that give these areas worldwide significance not only for unusual species but as a gene bank with untapped and even undiscovered resources for medicinal and other important human uses. They play an important part in stabilizing the ecosystem, particularly in the control of drainage, microclimate, water purification and soil formation. Coastal peat swamps act as a buffer between marine and freshwater systems, preventing excessive saline intrusion into coastal land and groundwater. Peat swamps often serve as a natural gene bank, preserving potentially useful varieties of plant species. At a global scale the peat swamp forests contribute to the storage of atmospheric carbon that is an agent of global warming, helping to slow down that process. Peat swamp forest areas can also be very productive through the managed extraction of fish, timber and other forest products. There is a widespread misapprehension that peat covered catchments can function as aquifers or giant sponges, absorbing and storing water during wet periods, preventing floods, and releasing water slowly during dry periods. However, since water tables are at or close to the level of the forest floor, and there is an already saturated upper peat layer, there is little further storage space available for additional water. The amount of water in peat depends on the level of the water table, which after prolonged dry periods naturally induces a low water table. Since lateral infiltration is facilitated by roots, peat will easily retain modest amounts of rainfall following dry spells. However, surplus water retained during the wet season is attributable to flat topography and poor drainage rather than any sponge-like function of the peat substrate. High water tables and waterlogged conditions within undisturbed peat swamp areas imply that further water-storage capacity is limited, even though peat soils have a high water-holding capacity. Peat Swamp water storage capacity is very low and drainage response occurs within 1-2 hours after rainfall has started. Even after prolonged dry periods (more than 10-15 days), storage capacity is only 20-30mm of rain; the role of predominantly rain-fed peat lands as storm flood protecting environments is restricted to the capacity of individual catchments (Peat Swamp Forest Project Malaysia). Peat lands receive water from surrounding areas. Peat deposit, due to its flatness, act as a buffer and delay the discharge of water (MacFarlane, 1969). Peat and its drainage are very acidic, poor in nutrients, and saturated with organic matter, making it unsuitable for potable use without extensive treatment. Poor drainability of peat soils makes it very difficult to retrieve a significant flow of water from a peat swamp area especially if attempting to tap supplies from sub-surface sources. Extraction of significant quantities of water would have serious impacts on the natural hydrological system, causing a lowering of the groundwater and, over time, leading to peat decay and subsidence.

Peat swamp forests play a very important role in maintaining a wider regional ecosystem balance and critically support social and economic systems through the functions they provide. Pearland is widespread in Sarawak, Malaysia occupying approximately 13.0 percent or 1.7 million hectares of the total land area (Hamsawi and Ikhwan 1996). Mukah Division, which covers an area of 6,997.6 square kilometres, is largely made up of low-lying peatland. Peat has been considered a problem soil because early attempts by some small holders to cultivate it for subsistence agriculture activities often failed, and farms were abandoned. Peat has good water holding capacity, however, and is able to hold large amounts of nutrients, which make it suitable for crop growth (Kanapathy, 1975). These favourable properties of peat have opened up opportunities for large-scale cultivation of crops, especially sago and oil palm in Mukah.

When a peat swamp area is flooded, the reduction in water velocity associated with it spreading over a wide area, together with the retarding effects of vegetation, allows suspended sediments to settle. Water flowing back into rivers will then be largely sediment free. Peat covers a significant portion of the plantation land of Malaysia.

There are about 2.84 million hectors of peat in Malaysia, accounting about 8% of the total land area of the country. Sarawak has the largest peat area of the country, which is about 1,765,547 ha, representing 62.25% of the total peat of the country and 13% of the land area of the state.

Table 1.1. Peat land product

Types of materials collected	Percent
Construction materials	61.5
Firewood	71.7
Bamboo	57.1
Rattan	35.1
Mushroom	60.0
Weaving materials	36.6
Wild vegetables	93.7
Others	8.3

Source: Peter et al.(2006).

Peat area in peninsular Malaysia is about 984,500 ha, representing 34.72% of its total peat area and 4.46% of the total land area of the country, while Sabah 86,000 ha, representing 3.03% of the total peat area and 1% of the total land of the state (Jamaluddin, 2002). About 313,600ha of peat area in Peninsular Malaysia have been developed for oil palm plantations which represent 32% of the total peat area of the country.

Plantation land of oil palm on peat increased over the years from 243,000 to 325,200 hectares, 366,48 to 437,782 hectares, and 17,200 to 25,800 hectares from 1998 to 1999 in Peninsular Malaysia, Sarawak and Sabah, respectively (Wong,1989).

1.2. Functions and Values of Peatlands

Peat swamp forests play a very important role in maintaining a wider regional ecosystem balance and critically support social and economic systems through the functions they provide. Table 1.2. presents an overview of the main peatland ecosystem values, categorized into direct uses, indirect uses and non-use values. These three types of values together form the Total Economic Value of a peatland.

Table 1.2. Services and goods provided by peat swamps

Peatland Values	Examples
Direct use (Production functions)	• Source of water • Recreation • Direct extractive use of biodiversity: 　• food (e.g. fish) 　• medicinal plants 　• ornamental plants 　• aquarium fish 　• timber 　• non-timber forest produce like rattan and other plants for construction purposes, fuel and handicrafts
Indirect use (Regulation functions)	• Storage and sequestration of carbon • Reduction of downstream flood peaks by absorbing floodwaters • Maintenance of base (minimum) flows in rivers by releasing water slowly during dry periods • Prevention of saline water intrusion by maintaining base flows and water table levels
Non-use	• Spiritual, historical and cultural values • Aesthetic values • Biodiversity attributes e.g. species richness and endemism

Source: ASEAN, 2006.

These values are the beneficial outcome of the hydrological, chemical and biological processes within the ecosystem. Hydrological functions are especially important at the local scale. The peat, acting as a sponge, absorbs water during wet periods and releases it slowly during dry periods.

Thus, intact peatlands have a very great potential to prevent loss of life and damage to infrastructure by reducing flooding downstream of the peatland, while the maintenance of minimum flows in rivers in the dry season can support irrigation works downstream and prevent saline water intrusion up rivers (ASEAN, 2007; www.ckpp.org).

Table 1.3. Engineering Significance of Peat

Engineering field	Purpose of involvement
Design engineering	Operation optimization Traficability analysis Vehicle-terrain interaction mechanization Vehicle design Air-cushion system Novel air-cushion protection system Suspension system Electrical and instrumentation
Agricultural engineering	Vehicle adaption for mechanization Equipment adaptation
Hydraulics	Drainage assessment Drainage control Drainage manipulation Peat exploitation Peat product manufacturing
Engineering research	Physical properties of peat Peat mechanics Prediction for utilization Environmental interpretation Optimization studies Environmental manipulation
Environmental impact assessment	Ecological impact of peat on agriculture Economic value of peat land Green House Emission (GHG)

1.3. Engineering Significance of Peat

Engineering treatment is required for access and construction in peat. Conservation of water, construction, foresty, drainage and agricultural applications are example of operations that are required special design. The construction and highway design engineers need to aware of the dynamic of the water regime in the peat to be crossed. Requirement of intensity and frequency of loading need to be considered in relation to total, year-round environment.

The mechanical engineers need to play important role on designing off-road-vehicle for the mechanization of 1.2: Engineering significance of peat swamp agricultural products and transportation of industrial products over the peat which must be encountered the limitation of peat traficability. There are examples where expensive vehicles and drainage machinery have been developed without proper consideration of the inherent, fixed design in the organic terrain. Immobilization, when it ensues in these cases, portrays failure of purpose, serious interruption in operations and perhaps more significantly inadequacy, if not irresponsibility, in engineering practice. In vehicle design, the limit of tolerance for ground pressure in relation to the terrain bearing capacity are known, whereas much still has to be learned about design for suspension systems. Engineering significance of peat swamp is given in the Table 1.3.

1.5. Ecological Impact of Agricultural on Peat Land

Tropical peat swamp forest destruction for reasons of agricultural land development leads not only to emission of greenhouse gases (GHG) but also to various serious regional environmental problems. One of the regional environmental problems caused by the destruction of tropical peat is the oxidation of Ferrite Sulfide (FeS_2) within the sediment underneath the peat layer in the coastal area of the tropical region, an effect much like sulfuric acid pollution in coal mining areas of Europe (Monterroso and Macias 1998; Balkenhol et al.2001). Then, sulfuric acid concentration increases in the soil, and a large proportion of the nutrients is lost. Discharged sulfuric acid from the soil causes the acidifi cation of river water and subsequent effects on the littoral zone. Indonesia has large areas of tropical peatlands, and Kalimantan has about 6.7 Mha of peatlands. The 1 million ha "Mega Rice Field Project, Central Kalimantan" started in 1995, and the project aimed to establish irrigated rice fi elds on 1 million ha of peat swamp forest. The project, however, stopped in 1997, and most of the fields have been abandoned (Akira Haraguchi, 2007).

Large quantities of carbon are stored in tropical peat lands. Estimates suggest that 5,800 tonnes of carbon per hectare can be stored in a 10-metre deep peat swamp compared to 300-500 tonnes per hectare for other types of tropical forest (UNDP, Malaysia 2006). Tropical peat lands, besides acting as stores of carbon, actively accumulate carbon in the form of peat. Because decomposition is incomplete, carbon is locked up in organic form in complex substances formed by incomplete decomposition. Drainage of peat swamps destroys this useful function and may contribute to global warming through the release of CO_2 into the atmosphere. Water table level depth in peat plays an important role in peat accumulation and decomposition dynamics, and thereby also in soil CO_2 emissions, all of which form important components in terrestrial carbon storage and soil-atmosphere greenhouse gas dynamics (Rieley and Page, 2005). Appropriate water management may improve vegetation growth rate on peat, and thus increase the potential for peat accumulation in comparison to decomposition carbon losses. The relationships of CO_2 emission to water table, peat moisture and humification were studied in a secondary peat swamp forest floor at the Vodoi National Park, Vietnam (Duong et al. 2006). Water table level is artificially maintained high and the forest floor is flooded for most of the year in order to reduce risks of fire. The lowest water table was at -30 cm below the peat surface and the moisture of peat above the water table level was at the lowest 70% w/w (from the 80% water holding capacity). At these water tables CO_2 release from the peat surface was directly proportional to the water table level. The rate of CO_2 release from peat, excluding plant root respiration, was significantly higher and not clearly dependent on peat moisture content in the less humified topmost 10 cm peat (litter) layer compared to the more humified deeper peat. The highest CO_2 release rate from decomposition and faunal activity was found in the topmost 10 cm litter layer at a moisture content of 67-85% w/w (from the 80% water holding capacity). The CO_2 release rate below the litter layer was less than 0.02 mg CO_2 g^{-1} h^{-1}, and this decreased significantly when the water content was < 40% of the water holding capacity. As the water table level in the forest floor never fell lower than -30 cm from the ground surface and the moisture of peat above the water table was always above 80% of water holding capacity, it was concluded that the near surface litter played the major role in CO_2 emission from peat in Vodoi National Park.

Tropical peat land forests seem to change in a for agricultural land such pady, sago, palm oil and so on and continue to be converted to meet the food demand of the growing population in many of Asian countries (Watson *et al.*, 2000). Peat land has a significant amount of carbon stock and some nitrogen in the soil which could be a source of the greenhouse gases carbon oxide (CO2), methane (CH4) and nitrous oxide (N2O). It is reported that the path of methane from paddy soil to the atmosphere is mainly through the rice plant (Inubushi *et al.,* 1989) and also the greenhouse gas emitting, and CH4 producing and oxidizing, soil microbial activities in a sago palm plantation in Sarawak, Malaysia (Inubushi *et al.,* 1998), paddy field, and upland field in South Kalimantan and Jambi, Indonesia (Inubushi *et al.,* 2003; 2005; Furukawa *et al.,* 2005) in order to give a direction for such land conversions.

Increase in CO_2 emission from peatland is usually preceded by deforestation and peatland fires. Subsequent land management systems determine the rate of emission. Oil palm requires drainage to about one metre depth and this leads to below ground CO_2 emission of about 91 t ha-1 yr-1. Some practices, for example, the 'sonor' system, which involve burning of surface peat for rice cultivation can lead to CO_2 emissions as high that those from oil palm plantations. Several other farming systems tolerate shallow or no drainage and thus have lower emissions. Peat land degradation and fires emit about 2,000 Mt CO_2 each year; 1,400 Mt of which is released from forest and peatland fire and the other 600 Mt is from peat degradation following removal of the forest (Hooijer *et al.* 2006).

1.5.1. Estimate of Net Emissions

Green house Gas emission (mainly CO_2) can be estimated by using the following equation:

$$GHG=(E_a-S_a)+(E_b-E_o) \qquad (1)$$

where, GHG is the green gas emission, E_a is the emission from above ground (mainly caused by biomass burning), S_a is Carbon sequestration into the above ground biomass, E_b Emission from below ground burning, E_o is the emission from below ground oxidation.

For calculating E_a, a production cycle is assumed as long as 25 yr, resembling that of oil palm. When a wetland (logged over) forest with C density of about 200 t ha^{-1} (Rahayu et al. 2005) is deforested, almost all of the carbon is likely to be emitted over the 25 yr period through burning and decomposing. The E_a is then 8 t C or about 29 t CO_2 ha^{-1} yr^{-1}. This value was applied for all evaluated land uses. S_a is assumed negligible in annual crop based systems. A mature oil palm plantation contains about 100 t C (Tomich, 1999), which corresponds to Sa of about 4 t C or 15 t CO_2 ha^{-1} yr^{-1}. S_a for rubber is assumed to be equal to oil palm. E_b is significant in some systems such as the 'sonor' system. Widespread fires in the wetland of South Sumatra and Lampung in the periods of 1991, 1994 and 1997/98 were partly a result of sonor; a traditional system of wetland rice cultivation. Sonor is practiced by using fire during prolonged droughts of five to six months (every four years). As the wetlands dry out, surface vegetation is burned and rice seeds are broadcast on the ash-enriched peat soil. This became more common as the incidence of droughts increased, new areas became accessible through canals, and new migrants also adopted the practice (Chokkalingam, et al., 2006). Since the peat is dry during the long dry season, fire can easily burn a layer of peat exceeding 10 cm. If it is assumed that 10 cm of surface peat burns in four years then 2.5 cm of peat is burned in one year under the sonor system. For annual upland farming, with once or twice a year burning, but with less biomass to burn, a burning depth of 1 cm per year is assumed (Famudin et al. 2008).

With peat bulk density ranging from 0.10 to 0.34 t m-3 and organic C content ranging from 0.3 to 0.5% (w/w) (Wahyunto et $al.,$ 2004), then carbon density in a one metre layer of peat ranges from 530 to 1260 t ha-1 or 900 t C ha-1, or 9 t ha^{-1} cm^{-1} on average. E_b from the sonor systems then is about 22.5 t C or 83 t CO_2 ha^{-1} yr-1. For annual upland crop it is about 33 t CO_2 ha^{-1} yr^{-1}. E_o estimates from different literature sources vary widely. Some show CO_2 emission as low as 4 t ha^{-1} yr^{-1} from paddy field with drainage depth of 10 cm and as high as 127 t ha^{-1} yr^{-1} from drained secondary forest with drainage depth of 38 cm. Despite the variation, Hooijer et $al.$ (2006) suggested CO_2 emission of about 0.91 t ha^{-1} yr^{-1} per cm drainage depths ranging from 25 to 110 cm. This relationship is used in this paper. Based on field observation and various literatures the drainage depths for sonor, annual upland crops, lowland rice, rubber and oil palm are 10, 20, 10, 20 and 100 cm, respectively. Land-use change from secondary peat swamp forest to paddy field in coastal peat land at Gambut tended to increase annual emissions of CO_2 and CH_4 to the atmosphere, while changing land use from secondary forest to upland tended to decrease these gas emissions.

Kazuyuki and Hadi (2008) studied on greenhouse gas emission in Jambi, Sumatra by changing land use from drained forest to lowland paddy field significantly decreased the CO_2 and N_2O fluxes, but increased the CH_4 flux in the soils. They reported that change from drained forest to cassava field significantly increased N_2O flux, but had no significant influence on CO_2 and CH_4 fluxes in the soils. Average CO_2 fluxes in the swamp forests of 94 mg C m^{-2} h^{-1} were estimated to be one-third of that in the drained forest. Groundwater levels of drained forest and upland crop fields had been lowered by drainage ditches while swamp forest and lowland paddy field were flooded, although groundwater levels were also affected by precipitation. Groundwater levels were negatively related to CO_2 flux but positively related to CH_4 flux. The peak of the N_2O flux was reported from their study at -20 cm of groundwater level.

1.5.2. In Situ CO₂ Measurement

Duong et al. (2006) measured the CO_2 emission during the dry season at five permanent points located in the peat area of Vodoi National Park, Vietnam. At each point, the depth of the water table level was recorded and peat moisture content above the water table was determined at 10 cm intervals up to the soil surface. For the measurement of the rate of CO_2 emission, a round aluminum chamber (with a diameter of 31 cm and a height of 12 cm) was inserted into the peat to a depth of 1 cm from its lower edge and the rate of CO_2 emission was calculated from concentration increase in the closed chamber detected by a EGM-4 infrared gas analyzer. The relationship between peat moisture and CO_2 flux from peat decomposition, without live roots, was measured at depths of 0-10cm, 10-20cm and 20-30cm. For this purpose, peat samples collected from each depth were placed in air-tight boxes of 160cm diameter and 10 cm height. The peat was packed in the boxes to a 5 cm thick layer. CO_2 respiration was measured by infrared gas analyzer after the peat material reached water holding capacities of, 80%, 60%, 40% and 20%, respectively.

The rate of CO_2 releases from peat – that is, from decomposition and faunal activity but excluding root respiration – was significantly higher in the less humified top 10 cm (litter layer) than in the more humified deeper peat layers, irrespective of the moisture content. The highest CO_2 release rate from peat decomposition and faunal activity was found in the top 10 cm litter layer at a moisture content of 67-85% w/w (from the water holding capacity of 80%). By contrast, the rate of CO_2 release via decomposition and faunal activity in more humified peat material below the litter layer was very low, being less than 0.02 mg CO_2 g^{-1} h^{-1} at water content ranging from 60% to 100% of the water holding capacity, $i.e.$ higher than 50% moisture (w/w). The CO_2 release rate was even less at moisture content <40% of the water holding capacity. As the water table level in the forest floor did not sunk deeper than - 30 cm below the soil surface and the peat moisture content above the water table was over 80% of the water holding capacity, it was conclude that the near surface litter layer played the major role in CO_2 emission from peat materials in Vodoi National Park.

1.6. Drainage of Peat Land

Drainage is needed to make peat swamp waterlogged lands suitable for agriculture or other land uses and avoiding flooding by the evacuation of excess rainfall within a certain period of time. To be able to feed the growing world population and to banish hunger, food production needs to be doubled in the next 25 years. In south-east Asia, where fertile land has become scarce, agriculture development more and more focuses on marginal soils such as peatlands, acid sulphate soils, and steep land. Peat lands in south-east Asia cover about 27.1 Mha, or about 10% of the total land area (Hooijer $et\ al.$, 2006). These peat lands, which are waterlogged most time of the year, are dome-shaped. It should be realized that because of the dome-shape topography, peat domes cannot be irrigated by gravity from the surrounding rivers. The only source of water is rainfall. The rainfall is not distributed evenly over the year (Ritzema and Wösten, 2002). The average dry season (monthly rainfall <100 mm) can last for 3 to 4 months. Under these conditions the water table depth can fall to one metre or more below the peat surface (Takahashi $et\ al.$, 2002). Without water conservation, this can lead to severe and persistent moisture deficits in the surface peat layer and thus to increased oxidation and risk of fire. Drainage is needed during the monsoon season to control the water table and to remove excess surface and subsurface water from the land. The water management requirements also vary within a year. Water table control is required the whole year around, with removal of excess water only during periods with excess rainfall, whilst water conservation is essential during prolonged dry periods. Water conservation is needed to control the water table at a higher level throughout the year to avoid excessive subsidence and to reduce the risk of fire.

Peat structures are needed to control the drainage as it is so permeable. The dynamic storage capacity in the drainage system is small compared to the recharge by excess rainfall and the corresponding discharge. Therefore it is possible to use the steady-state approach for the design (Beekman, 2006). Structures act as barriers to prevent the flow but water cannot be stored for long periods as it will seep away through the surrounding peat. Compared to mineral soils, peat has a much higher infiltration capacity, drainable pore space and hydraulic conductivity, but a lower capillary rise, bulk density and plant-available water (Wösten *et al.*, 2003). Another major difference is the subsidence behaviour of peat; it is partly caused by oxidation and is never-ending. Oxidation leads to CO_2 emission, which under Borneo conditions, is estimated to be in the order of 26 tonnes per hectare per year (Ritzema et al., 2006). In addition to the loss of peat by oxidation, the excessive subsidence rates result in a pronounced drop in the elevation of the land reducing the efficiency of the drainage system.

The majority of the lowland tropical peat swamps are completely rain-fed. Water flow from upland areas does not enter them. The rainfall either evaporates or is transported from the swamps as near surface run-off, inter-flow, or groundwater flow. Under unsaturated conditions, almost all of the rainfall infiltrates owing to the high hydraulic conductivity (up to 30 m per day) of the surface (approximately first 1 metre) peat layer (Takahashi *et al.*, 2002). The high hydraulic conductivity, combined with a relatively low hydraulic gradient in the relatively flat peatlands, results in a rapid discharge of rainfall through the top peat layer towards the surrounding rivers or drainage channels. The general water balance of a peat swamp can be expressed as follows:

$$L_{rain} = E_{evaporatio\ n} + Q_f + \Delta L$$

(1.2)

where, L_{rain} is the total rainfall (m), $E_{evaporatio\ n} = f(T)$, $E_{evaporation}$ is the total evaporation in m and T is the temperature in degree Celsius, $Q_f = V \times A_{drain}$, where Q_f is the flow rate in m^3s^{-1} and ΔL is the change of storage in m.

REFERENCES

Diana Nurani, Sih Parmiyatni, Heru Purwanta, Gatyo Angkoso, Koesnandar. 2006. Increase in ph of peat soil by microbial treatment.

Fauzi Y., Widyastuti, Y.E., Satyawibawa, I. and Hartono, R. (2006). *Cultivation, product utilization and market analysis of oil palm.* Penebar Swadaya, Jakarta. (In Indonesian).

Hamsawi, A. O. A., and Ikhwan, A. H. (1996). *Prospect in the development of peat for oil palm in Sarawak.* Paper presented at the 1996 Seminar on Prospect of Oil Palm Planting on Peat in Sarawak: The Golden Opportunity, 18-19 March, Sibu, Sarawak.

Hooijer, A., Silvius, M., Wösten, H. and Page, S.E. (2006) PEAT-CO2, Assessment of CO2 emissions from drained peatlands in SE Asia. Delft Hydraulics report Q3943.

Hooijer, A., Silvius, M., Wösten, H., and Page, S.E. (2006) Peat-CO2, Assessment of CO2 Emissions from drained peatlands in SE Asia. *Delft Hydraulics report Q3943.*

Inubushi, K., Furukawa, Y., Hadi, A., Purnomo, E. and Tsuruta, H. (2003) Seasonal changes of CO2, CH4 and N2O fluxes in relation to land-use change in tropical peatlands located in coastal area of south Kalimantan, *Chemosphere*, 52, 603-608.

Kanapathy, K. (1975). Growing crops on peat soils. *Soil and Analytical Services, Bull. No.4,* Ministry of Agriculture, Kuala Lumpur.

Kazuyuki Inubushi1 and Abdul Hadi (2008). Effect ffect of land-use management on greenhouse gas emissions from tropical peatlands.

Koesnandar, Parmiyatni, S., Nurani, D., Wahyono, E. (2006). Government Role on Research and Application of Technology for Peatland Utilization. *National Seminar on peatlands and their problems.* March 21st 2006. University of Tanjungpura, Pontianak.(In Indonesian).

Laegreid, M., Bockman, O.C. and Kaarstad, O. (1999). *Agriculture, Fertilizers and the Environment.* Norsk Hydro ASA: CABI Publishing.

Lusiana, B., van Noordwijk, M., and Rahayu, S. (eds.) *Carbon Stock Monitoring in Nunukan, East Kalimantan: A Spatial and Modelling Approach.* World Agroforestry Centre, SE Asia, Bogor, Indonesia.

Page, S.E. and Rieley, J.O. (1998) Tropical peatlands: a review of their natural resource functions with particular reference to Southeast Asia. *International Peat Journal.* 8. 95-106.

Rahayu, S., Lusiana, B., and van Noordwijk, M. (2005) Above ground carbon stock assessment for various land use systems in Nunukan, East Kalimantan. pp. 21-34. In:

Rieley, J.O., Ahmad-Shah, A.A. and Brady, M.A. (1996) The extent and nature of tropical peat swamps. In: E. Maltby. C.P.Immirzi and R.J. Safford (eds.) *Tropical Lowland Peatlands of Southeast Asia.* IUCN. Gland. pp.17-53.

Ritzema, H.P. and Wösten, J.H.M. (2002) Water Management: The Key for the Agricultural Development of the Lowland Peat Swamps of Borneo. *Proceedings International Workshop on Sustainable Development of Tidal Areas,* July 22, 2002, ICID, Montreal, pp. 101-113.

Ritzema, H.P., Hassan, A.M.M. and Moens, R.P. (1998) A new approach to water management of tropical peatlands: a case study from Malaysia. *Irrigation and Drainage Systems* 12,123–139.

Suyanto, Ohtsuki, T., Ichiyazaki, S., Sadaharu, Subroto, A., Koesnandar and Mimura, A. (2003). Possibility of efficient composting on palm oil mill fibre by optimized solid state fermentation using thermophilic fungus *Chaetomium* sp., *Journal of Microbiology Indonesia,* Vol. 8 No. 2:57-62.

Takahashi, H., Shimada, S., Ibie, B.I., Usup, A., Yudha and Limin, S.H. (2002). Annual changes of water balance and a drought index in a tropical peat swamp forest of Central Kalimantan, Indonesia. In: Rieley, J.O., Page, S.E. with Setiadi, B. (eds), *Peatlands for People: Natural Resource Functions and Sustainable Management, Proceedings of the International Symposium on Tropical Peatland,* 22-23 August 2001, Jakarta, BPPT and Indonesian Peat Association, pp. 63-67.

Tomich, T.P. (1999) Agricultural Intensification, Deforestation, and the Environment: Assessing Trade offs in Sumatra, Indonesia. *Proceedings of the International Symposium "Land Use Change and Forest Management for Mitigation of Disaster and Impact of Climate Change",* Bogor, 19-20 October, 1999. pp. 1-27.

Wösten, J.H.M., Ritzema, H.P., Chong, T.K.F. and Liong, T.Y. (2003) Potentials for peatland development. In: Chew, Dr. D. and Sim Ah Hua (Ed.). *Integrated Peatland Management for sustainable development. A Compilation of Seminar Papers.*Sarawak Development Institute, Sarawak, Malaysia, pp: 233-242.

Chapter 2

PRODUCTIVITY AND ECONOMIC VALUE OF PEAT

2.1. INTRODUCTION

Many poor and vulnerable peat-dependent communities are living in the Berbak - Sembilang area. The challenges and opportunities facing these communities flow from arange of ecological, socio-economic, market and other forces, many of which are beyond the direct control of the communities. An understanding of this context highlights the interaction between the micro, meso and macro environments and the importance of ensuring that any initiative focusing on poverty reduction makes links between these different levels. The vulnerability context considers the external environment within which individuals and communities live, and over which they have very limited control. In the Berbak-Sembilang area this is shaped often by interrelated environmental and socio-economic factors. Periodic floods and droughts are a regular part of the seasonal cycle for communities in the project area. There are strong signs, however, that these natural shocks are being exacerbated by human activities such as illegal logging and land conversion for plantations and agriculture. As explained above, the hydrology of forested peatland ecosystems is highly sensitive to both localized and up-stream disturbances. For this reason, efforts to mitigate peatland degradation and the associated loss of critical environmental services, have to focus not only on the local situation, but also must engage at the broader spatial and policy levels. These events of the last decade mark a radical shift in the vulnerability of peat swamp forests, which do not burn under natural conditions. The very real danger is that fire in the peat swamp forests will cause irreversible damage to the ecosystem, which will have radical effects on down stream communities, even those at a great distance from the burnt area. Once peat has been lost, it will not return, and if the damage is severe enough, the natural forest will not rejuvenate. This will result in the loss of the environmental services mentioned above, with associated negative impacts on local community livelihoods.

2.2. OPERATION OF PEAT LAND

Illegal logging is widespread in both Berbak National Park and the MKPSF, which lies in the buffer zone of Sembilang National Park. This logging contributes to the degradation of the peat and forest through a complex array of interactions. Felling of trees changes the local conditions and forest cover, while at the same time leaving dry fuel for fires. Additionally, in some areas loggers dig ditches through the forest in the peat in order to facilitate removal of the logs.

This is especially prevalent in the MKPSF, where recent studies identified more than 100 of these ditches. The ditches drain the peat, leaving it more vulnerable to fire. With the physical conditions modified in ways that predispose the area to fire, the presence of the loggers, with cigarettes and cooking fires, greatly increases the risk of fire.

Land conversion in the upper watersheds of the Air Hitam Laut and Merang Rivers (which run respectively through Berbak National Park and the MKPSF) poses a high threat to the target areas and population. As part of a comprehensive study recently completed by Wetlands International and various research institutions from the Netherlands, a hydrological model was used to determine the impacts of the very real possibility of expansion of oil palm plantations in the upstream area of the Air Hitam Laut river basin. The study concluded that "the consequences for the coastal agricultural areas will be a severely reduced water discharge by the river in those (dry) periods when it is most needed for agricultural production (Silvius, 2005). Expansion of oil palm plantations upstream will thus significantly impact on agricultural production in the downstream area" (Silvius, 2005). Livelihood assets (or capital) are the main building blocks upon which peoples' livelihoods are built.

This section is intended to provide a general snapshot of the livelihood assets of the target population to assist in identifying opportunities for strengthening those assets in order to enhance livelihood options, thus reducing poverty. As WI-IP and its partners will not be able to build on all the opportunities and address each gap under the project, this snapshot will also help identify areas for close collaboration with other organizations and institutions.

The target communities are generally quite remote, with some only being accessible by water. Roads, where they exist, are usually of very poor quality and impassable at certain times of the year.

This isolation creates very specific challenges, and the physical capital of the communities tends to be low. For example, in Air Hitam Laut village there is no market. Lack of potable water is frequently cited as one of the main problems facing community members. Electricity is only available for a few hours each day.

2.3. COMMUNITIES INVOLVEMENT

Education levels tend to be low in the target communities. There is difficulty accessing up-to-date information about agricultural techniques, and government extension services rarely reach the communities. As a result, people's farming skills tend to be basic, and there is little knowledge of new developments and alternative techniques. In a number of consultation meetings in the target and other similar communities, a lack of financial resources was cited as one of the biggest obstacles to developing income generating activities.

The lack of options available to the population has a number of negative effects, including, among others, out-migration, unsustainable exploitation of natural resources, people entering into a cycle of debt, for which it is difficult to extricate themselves and involvement in the dangerous work of illegal logging, at the mercy of often unscrupulous "logging bosses" (*cukong*). Historically the Berbak-Sembilang area has been well endowed with natural resources that provided a number of benefits for local communities.

For the mainly small-scale agriculture and fisheries dependent people in the area, the degradation of these natural resources over recent decades as a result of forces such as uncontrolled logging, over fishing and land conversion, has severely eroded people's natural capital.

With increasing risk of fire and the trend towards even more severe degradation of the peatland ecosystem this erosion of natural capital is likely to continue.

2.4. UTILIZATION OF PEATLAND

Peatland is widespread in Sarawak, Malaysia occupying approximately 13.0 percent or 1.7 million hectares of the total land area (Othman et. al, 1996). Mukah Division, which covers an area of 6,997.6 square kilometres, is largely made up of low-lying peatland. Peat has been considered a problem soil because early attempts by some small holders to cultivate it for subsistence agriculture activities often failed, and farms were abandoned (Lim *et al.*, n. d.). Peat has good water holding capacity, however, and is able to hold large amounts of nutrients, which make it suitable for crop growth (Kanapathy, 1975). These favourable properties of peat have opened up opportunities for large-scale cultivation of crops, especially sago and oil palm in Mukah.

Acknowledging that sago has been Mukah's niche agricultural-based product, the Land Custody Development Authority (LCDA) allocated 30,000 hectares of peatland for sago cultivation in the Mukah watershed. From 1987 to 1994, LCDA has used approximately 8,000 hectares of deep peat in Mukah Division for sago (Kueh, n. d), by establishing a number of sago plantations (the Mukah, Dalat and Sebakong Plantations). LCDA's Crop Research and Application Unit (CRAUN) has also been conducting downstream research to determine the potential of sago for the pharmaceutical, food, chemical, and cosmetics industries.

2.4.1. Sago Cultivation

Sago has been cultivated by the local communities for generations. Apart from being their major source of income and sustenance, sago cultivation has become part of their daily life. The size of land cultivated with sago ranged from less than 1.0 hectare to more than 3.0 hectares, with a mean size of 2.3 hectares (Peter et al.,2006).

2.4.2. Utilization of Peat Forest

Peatland in the Mukah watershed was utilized mainly for small scale agriculture and plantation purposes. Peat forest, however, also served various uses for local communities who derived a range of different products from it, including food, construction materials, firewood, weaving materials, rattan and others.

2.4.3. Utilization of Rivers

A network of rivers traverses the Mukah watershed and are used by the local communities for various purposes, mostly transportation (96.0%), specifically, transportation of sago logs (85.0%) (Peter et al.2006). Other uses of rivers mentioned by the respondents were for fishing, recreation and domestic purposes.

REFERENCES

Kanapathy, K. (1975). Growing crops on peat soils. *Soil and Analytical Services, Bull. No.4*, Ministry of Agriculture, Kuala Lumpur.

Kueh, H. S. (n. d.). *The NPK requirement of the sago palm (Metroxylon sagu Rottb) on anundrained deep peat soil in Sarawak.*

Lim, E. T., Ahmad, B., Tie, T. L., Kueh, H. S., and Jong, F. S. (n. d.). *Utilization of tropical peats for the cultivation of sago palm (Metroxylon spp.).*

Peter Songan, Gabriel Tonga Noweg, Wan Sulaiman Wan Harun and Murtedza Mohamad. 2006. Sustainable livelihood of peatland dwellers in the Mukah watershed, Sarwak, Malaysia.

Silvius, M.J., H.W. Simons, and W.J.M. Verheugt, 1984. Soils, vegetation, fauna and nature conservation of the Berbak Game reserve, Sumatra, Indonesia. RIN Contributions to Research on Management of Natural Resources, xv + 146.

Chapter 3

ECONOMIC VALUE OF PEAT LAND

3.1. INTRODUCTION

The economic approach to valuing environmental changes is based on people's preferences for changes in the state of their environment. Environmental resources typically provide goods and services for which there are either no apparent markets or very imperfect markets, but which nevertheless can be important influences on people's well being. Examples include the quality of air, which affects people's health, crop yields, damage to buildings, and acidification of forests and fresh waters. However, the lack of markets for these services means that unlike man-made products, they are not priced; therefore their monetary values to people cannot be readily observed. The underlying principle for economic valuation of environmental resources, just as for man-made products, is that people' willingness to pay (WTP) for an environmental benefit, or conversely, their willingness to accept (WTA) for environmental degradation, is the appropriate basis for valuation. If these quantities can be measured, then economic valuation allows environmental impacts to be compared on the same basis as financial costs and benefits of the different scenarios for environmental pollution control. This then permits an evaluation of the net social costs and benefits of each scenario for the different environmental issue.

The monetary measure of the change in society's well being due to a change in environmental assets or quality is called the total economic value (TEV) of the change. To account for the fact that a given environmental resource provides a variety of services to society, TEV can be disaggregated to consider the effects of changes on all aspects of well-being influenced by the existence of the resource. TEV can be divided into use value and non-use value. The latter also being called 'passive use values'. Use value can be categorized as following (Perace and Turner, 1990):

- Direct use values, where individuals make actual use of a resource for either commercial purposes (e.g. - harvesting timber from a forest) or recreation (e.g. - swimming in a lake).
- Indirect use values, where society benefits from ecosystem functions (for example, watershed protection or carbon sequestration by forests)
- Option values, where individuals are willing to pay for the option of using a resource in the future (for example, future visits to a wilderness area).

- Non perceived values, where individuals are asked to evaluate a natural resource (for example, an individual may not be aware of the value of trees on green house effects and the importance of the latter.

Non-use value includes the following value (Greenley, Walsh and Young, 1981)

- Option value, a sort of insurance premium to ensure the supply of the environmental good, was thought to be positive for risk averse individuals facing supply uncertainty (Weisbord, 1964)
- Existence values, which reflect the fact that people value resources for 'moral' or 'altruistic' reasons, unrelated to current or future use.
- Bequest values, which measure people's willingness to pay to ensure their heirs will be able to use a resource in the future

However, given demand uncertainty, option value may be negative. Option value is only definitely positive when there is no demand uncertainty and where supply uncertainty can be completely removed by the proposed resource allocation. Bequest and existence value first suggested by Krutilla (1967) and may arise from either selfish or altruistic motives. Their existence is suggested by the public's WTP to preserve, for example, the blue whale and rain forests. Typically it is not possible to separate existence and bequest values. To arrive at an estimate of the net change in societal well-being arising from an environmental change, we must consider each of these elements in turn. The total economic value (TEV) of a change is the sum of both use and non-use values:

TEV = use values + non-use values = Non perceived + direct use +
+ indirect use + option + existence + bequest values

3.2. ECONOMIC VALUE OF PEAT SWAMP FORESTS

Peat swamp forests play a very important role in maintaining a wider regional ecosystem balance and critically support social and economic systems through the functions they provide. Table 3.1 presents an overview of the main peatland ecosystem values, categorized into direct uses, indirect uses and non-use values. These three types of values together form the Total Economic Value of a peatland. These values are the beneficial outcome of the hydrological, chemical and biological processes within the ecosystem. Hydrological functions are especially important at the local scale. The peat, acting as a sponge, absorbs water during wet periods and releases it slowly during dry periods. Thus, intact peatlands have a very great potential to prevent loss of life and damage to infrastructure by reducing flooding downstream of the peatland, while the maintenance of minimum flows in rivers in the dry season can support irrigation works downstream and prevent saline water intrusion up rivers (ASEAN, 2007; www.ckpp.org).v. Peat-swamp forests are characterized by a very high endemism.

They are one of the most important remaining habitats for several rare and endangered species. The forests of Borneo are especially known as the last wild habitat of orangutans.

The value of biodiversity is also important for medicine (ASEAN, 2006). A study by Kumari (1995, 1996) of the peat swamp forests of in Peninsular Malaysia, analyses the various benefits of moving from an existing unsustainable timber management system (base option) to sustained forest management overall. Kumari concludes that adopting more sustainable methods of timber extraction from peat swamp forest is preferable in economic terms. Although shifting to a sustainable harvesting system reduces the net benefits of timber harvesting, the case study suggests that this is more than offset by increased non-market benefits, primarily hydrological and carbon storage values.

The study evaluates the total economic value (TEV) of four options (one "unsustainable", three "sustainable") for logging a peat swamp forest in North Selangor. All three sustainable options have higher net present values than the unsustainable option, for which a TEV of about US$4,000 (M$10,238) per hectare was calculated. Over 90% of TEV in all cases is made up of timber and carbon storage benefits. Economic values considered include:

- direct use values associated with extraction of timber and Non-Timber-Forest Products (rattan and bamboo);
- indirect and direct use values associated with forest water regulation/purification services;
- direct use values associated with forest recreational benefits;
- indirect use values associated with forest carbon sequestration; and the existence and option values associated with wildlife conservation.

Peatlands are of considerable value to human societies due to the wide range of goods and services they provide. Peatlands help to maintain food and other resources and have functional significance far beyond their actual geographical extent. Now, the benefits provided has been discussed in general.

Table 3.1. Summary Results on Economic Value of Peat Swamp Forest in Malaysia and Cost for Change from Unsustainable to Sustainable Forest Management

Forest Good/service	Base Option (unsustainable taxcavator and canal (M$/ha)	Percent of Total Economic Value (TEV)	Change in TEV from base option to sustainable		
			sustainable taxcavator and canal (M$/ha)	Sustainable taxcavator and winch(M$/ha)	Sustainable Winch and Tramline(M$/ha)
Timber	2.149	21.3	-696	-399	-873
Hydrological - Agricultural	319	3.1	0	411	680
Wildlife Conservation	454	4.4	35	20	44
Carbon Sequestration	7,080	69.2	969	1,597	1,597
Rattan	22	0.2	88	172	192
Bamboo	98	1.0	0	-20	-20
Recreation	57	0.6	0	0	0
Domestic water	30	0.3	0	0	0
Fish	29	0.3	0	0	0
TEV	10,238	100	396	1,782	1,620

Source: Kumari, 1995.

3.3. DIRECT USE VALUE OF PEAT LAND

Direct use values, where individuals make actual use of a resource for either commercial purposes. The Direct use value of peat land includes the irrigation water for agricultural production, food and habitat for biodiversity, timber, recreation, storage and sequestration of carbon. The direct use of peat land has been described below.

3.3.1. Agricultural Production Functions

In regions where catchment areas are largely covered by peatlands, as well as in drier regions where peatlands indicate a rare but steady availability of water, they can play an important role in maintaining water supplies for drinking and irrigation water.

Table 3.2. Services and goods provided by peat swamps

Peatland Values	Examples

Direct use (Production functions)	• Source of water
	 • Recreation
	 • Direct extractive use of biodiversity:
	    • food (e.g. fish)
	    • medicinal plants
	    • ornamental plants
	    • aquarium fish
	    • timber
	    • non-timber forest produce like rattan and other plants for construction purposes, fuel and handicrafts

Indirect use (Regulation functions)	• Storage and sequestration of carbon • Reduction of downstream flood peaks by absorbing floodwaters • Maintenance of base (minimum) flows in rivers by releasing water slowly during dry periods • Prevention of saline water intrusion by maintaining base flows and water table levels
Non-use	• Spiritual, historical and cultural values • Aesthetic values • Biodiversity attributes e.g. species richness and endemism

Source: ASEAN, 2006.

The capacity of peatlands for agricultural production is generally low in the absence of intensive management (e.g. drainage, fertilization). In their natural state, peatlands have only marginal agricultural capability (Melling 1999, Rieley and Page 1997), thus restricting their use.

Important characteristics that inhibit agriculture are the very high groundwater table, the low bulk density and bearing capacity, the high acidity, the low availability of nutrients, and their subsidence upon drainage. Conventional agriculture involves drainage, fertilizing, tilling, compaction and subsidence, which eventually cut short the sustainability of peatland agriculture (Succow and Joosten 2001).

Agricultural development of tropical peatlands in South-east Asia only started a few decades ago. Table 3.2 shows that In Europe Hungry, Netherlans and Germany used the highest percentage of peatland for agriculture. From Asia, Malaysia and Indonesia used 45 and 25 percentage of peat land for agriculture respectively.

The agricultural successes are mainly due to the qualities of the surfacing sub-soil. As a result of continuous land hunger however, even the deeper peatland areas have become the target of agricultural development. Only a few commercial crops grow well on peatlands, including pineapple and oil palm.

More recently the dryland species *Aloe vera* has been introduced in Indonesia to the desiccated peatlands and is falsely propagated as a "sustainable" crop. New commercial and sustainable crops may include the indigenous Jelutung tree (*Dyera* sp., also known as the chewing gum tree, as it produces the latex that is used in chewing gum), which can grow under non-drained conditions. The development of fishponds in closed drainage canals offers interesting commercial potential for local communities. In addition, many tropical blackwater fish species are of interest to the aquarium industry.

Table 3.3. Peatland used for agriculture in selected countries

Area	Peatland Used for Agriculture(Km²)	% of total Peatland
Europe	124490	14
• Russia	70400	12
• Germany	12000	85
• Poland	7620	70
• Belarus	9631	40
• Hungary	975	98
• Netherlands	2000	85
USA	21000	10
Indonesia	60000	25
Malaysia	11000	45

Source: Joosten and Clarke 2002, Hooijer et al. 2006, JRC 2003.

New development opportunities are very much needed as poverty levels in Indonesian peatlands are generally higher than in non-peatland areas. In Malaysia Peatlands are distributed in the states of Selangor, Johor, Perak, Pahang, Sabah and Sarawak totalling to an estimation of 2.13 million ha, with the largest coverage of more than 1 million ha in Sarawak. Approximately 42% of peat land is utilized for agriculture and 52% of the total peat land area in Malaysia is designated as forest land. In Sarawak, 30% of total peat land has been used for agricultural production whereas in Peninsular Malaysia 32% of total peat land has been used for agricultural production as shown in Table 3.4.

Table 3.4. Distribution of Peat Land in Malaysia

Area	ha	%
Forest	387,642	52
Agriculture	312,292	42
Grass	22,061	3
Built-up	16,861	2
Water body	2,846	0.5
Mining(tin)	2,759	0.5

Source: Department of Environment, 2006.

Table 3.5. Distribution of Agricultural Crop Production in Malaysia in Peat Land

Crops	Area(ha)
Oil Palm	222,057
Rubber	39,082
Mixed Farming	13,173
Coconut	10,591
Pineapple	6,766
Mixed Horticulture	6,451
Rice	6,315
Orchard	5,244
Others	2,613
Total	312,292

Source: Department of Environment, 2006.

In Peninsular Malaysia, the North Selangor Peat Swamp Forest provides a significant supply of water (especially at the beginning of the dry season) to the adjacent Sekincan Rice fields - one of the key granary areas for the country.

Different types of agricultural crops have been produced in peat land in Malaysia which has been shown in Table 3.5. Among the different crops, palm oil constitutes the highest percentage of total agricultural production in peat land. Approximately 14% of European peatlands are currently used for agriculture, mainly as meadows and pastures. Also in North America, extensive areas of peatlands are cultivated for agriculture, as pastures and for sugar cane, rice, vegetables and grass sods. The commercial production of cranberries on peatlands in North America is a major business enterprise with the production of 6 million barrels. Large-scale cultivation in Southeast Asia is largely for estate crops (mainly palm oil, coconut and some sago) and rice. Sarawak is now the world's largest exporter of sago, exporting annually about 25,000 to 40,000 tonnes of sago products. Indonesia and Malaysia are the world's largest palm oil exporters, each producing about 43% of the global production (Basiron 2007). Recent poverty-induced agricultural encroachment has left over 80% of South Africa's coastal peat swamp forests denuded of its original vegetation. Communities have nowhere else to go and after the soil nutrients have been depleted, they move on to the next patch of remaining peat swamp forest (Marcel Silvius, pers.obs.). Agricultural development is also taking place in the high mountain peatlands of the Andean Paramos at over 3000 m altitude. Also in these "high mountain water towers", agriculture goes hand in hand with drainage and fires, and the practices are clearly not sustainable. The resulting decreasing water retention capacity may jeopardise the water supply to agricultural communities and cities downstream (Bermudes *et al.* 2000, Hofstede *et al.* 2002).

3.3.2. Direct Extractive Use of Biodiversity in Peat Land

Peatlands help to maintain food and other resources and have functional significance far beyond their actual geographical extent. Fur-bearers such as coyote, racoon, mink and lynx, and game species such as grouse, ducks, geese and moose, are often found in peatlands. In North America, black bears, hunted for food, fur, and traditional medicine (bladders), are also frequently found in peatlands. Wild reindeer (caribou in North America) are hunted for meat for local markets as well as for subsistence. An estimated 250,000 people in the Eurasian

Arctic depend on reindeer as a major food source. Caribou meat and hides are marketed in Canada on a small scale and in both Alaska and Canada caribou are hunted recreationally, generating income for guides, outfitters, and the service industry. For animal species that do not directly depend on peatlands, the habitat may contribute substantially to their continued presence in populated regions where few areas other than peatlands provide safe havens away from direct human disturbance. Peatland waters harbour many fish species, which in some regions, in addition to providing an important protein source to local communities, can be an object of sports fishing, generating income through the sales of fishing equipment and licences. Tropical black water fish diversity is extremely high. Black water species are attractive to sport fishing and the often very colourful species are attractive for the aquarium industry. Many economically valuable species are found in peat swamp forests. High quality timbers include Ramin (*Gonystylus bancanus*) and Meranti (*Shorea* spp.). Jelutung (*Dyera costulata*) is used for timber, pencils and for extraction of latex which is used in the production of chewing gum. Export values of wood products from peat swamp forests in Sarawak were over RM150 million in 1973 accounting for 60% of wood export revenue at the time. FAO (1974) estimated that a sustainable harvest with an annual export value of over RM200 million/annum from peat swamp forests would be possible. However, the amount of Ramin and other timber species harvested from peat swamp forests in the state has subsequently declined due to earlier heavy harvesting, poor regeneration and also conversion of peat swamp forests to agricultural land. Of the non-timber forest products rattans (*Calamus* spp.) latex (from trees of *Dyera* spp.) and incense bark (from the gemur tree, *Alseodaphne coriacea*) are important. These forests also contain a variety of medicinal plants such as *Cratoxylum arborescens* (for chicken pox), *Eugenia paradixa* (for diarrhoea) or *Piper arborescens* (for rheumatism) (Chai *et al.* 1989) . One of the important ornamental plants is the Pinang Rajah or sealing wax palm (*Cyrtostachys lakka*). Other species of commercial value from Peat swamp forests include fish and prawns with many species of ornamental fish being found in the black waters of peat swamp forests. Zakaria Ismail (1999) records that over 10,000 specimens of one species of ornamental fish (*Pseudobagrus ornatus*) were collected in the Nenasi peatlands in Pahang in 1997.

Southeast Asia peat swamp forest vegetation has been recognised as an important reservoir of plant diversity (Whitmore, 1984). Peat swamp forest has a relatively high diversity of tree species. For instance in Indonesia, more than 300 tree species have been recorded in swamp forests of Sumatra, some of which are becoming increasingly rare. 160 tree species have been recorded in Berbak National Park in Jambi province. (Giesen, 1991). In Malaysia, 132 tree species have been recorded from a 5ha plot in Bebar Forest Reserve in Pahang (Shamsudin, 1995) and 107 tree species have been recorded from North Selangor Peat Swamp Forest (Appanah *et al*, 1999). In Sarawak 242 tree species were recorded in peat swamp forests by Anderson (1963). While in Thailand, some 470 species of plant have been identified in the Narathiwat Peat Swamp Forest (Urapeepatanapong, 1996). Many of the plants are restricted or endemic to this habitat – for example 75% of the tree species found in peat swamp forest in Peninsular Malaysia is not found in other habitats and some of these species have relatively restricted distribution (Shamsudin, 1995). Peat swamp forests are home to a number of rare and endangered mammals such as Sumatran tiger (*Panthera tigris sumatranus*), tapir (*Tapirus indicus*), Asian elephant (*Elephas maximus sumatrensis*), lesser one-horned rhino (*Rhinoceros sondaicus*) and orang utan (*Pongo pygmaeus*). Recent studies in Kalimantan have indicated that peat swamp forest is one of the last strongholds for orang utan (Meijaard, 1995; Rieley, pers com 2001). Peat swamp forest also supports a diverse bird community. Prentice and Aikanathan (1989) recorded 173 species of bird in North Selangor Peat Swamp Forest of which 145 were breeding residents. Birds present include endangered species such as hornbills and the short toed coucal. Black-water rivers (peatland rivers) are important fish habitats that often have a higher degree of localized endemism than other rivers, and are important source of aquarium fishes. More than 100 species have been recorded from the North Selangor peat swamp forest (Ng et al, 1992). About 50% of these species are confined to black water rivers (including four new species to science described from this site in the early 1990's) while the other half are also found in other waters. In the black waters of Danau Sentarum in West Kalimantan more than 25 new species of fish have been described in the past 10 years.

3.3.3. Storage and Sequestration of Carbon

Peatlands are acknowledged to be important for sequestering and storing large amounts of carbon. Malaysia has the second largest extent of peatlands in Southeast Asia after Indonesia, which most are still intact thus contributing to sequestering carbon from the atmosphere as well as storage of large amount of carbon. Peatlands hold and sequester significant quantities of the world's carbon. Although peatlands cover only about 3% of the earth's surface, the total amount of carbon in standing vegetation and peat soil has been estimated at between 20-35% of the total terrestrial carbon (IGBP 1999/Patterson, 1999). Peatlands cover less than half of the area of tropical rainforests, however they contain three and a half times more carbon. Table 3.6 summarizes data from the German Advisory Council on Climate Change (GACCC, 1998) on carbon stocks and flows in peatlands.

It has been estimated that northern peatlands alone contain more than 500,000 million tonnes of carbon and that carbon sequestration in such peatlands over the last 5,000 years, at a rate of about 100 million tonnes/year, equals 100 years of fossil fuel consumption and represents a reduction in atmospheric CO_2 concentration of about 40 ppm (Gorham, 1991). Tropical peatlands although they only comprise about 10% of the peat area in the world have been accumulating carbon at a much faster rate than temperate peatlands. They also have large stores with about 5800 tonnes of carbon per ha stored in a 10 m deep peat swamp as opposed to 300-800 tonnes/ha for other tropical forests. Neuzil (1995) estimates that tropical peatlands store carbon at 3-6 times faster than in the temperate zone and so tropical peat deposits represent 25-40% of the annual global carbon storage in peatlands. Hence the loss or degradation of tropical peatlands may have a disproportionate impact on peatland carbon storage and the emission of greenhouse gasses. Concerns have been raised about the production of methane by wetlands, including peatlands, offsetting their role as carbon sequestering systems. However, many peatlands produce much less methane than other wetland systems (according to IGBP 1998, they produce only 20% of the methane produced by shallow water wetlands).

In addition, processes vary at different levels with a peat deposit. The lower levels of peat (catotelm) produce methane while the upper levels (acrotelm) at least partially oxidize methane released from the lower levels. The output of methane is determined by the production of methane by methanogenic bacteria and its removal by methanotrophic bacteria (Brown, 1998).

Table 3.6. Carbon stocks and flows of peat land

	Carbon Stores t C/ha Soil	Biomass t C/ha	Carbon Absorption t C/ha/yr
Global	1181-1537	150	1.0-0.35
Tropics	1700-3000	300-500	0.86-1.45
Boreal/Temperate	1314-1315	120	0.17-0.29

Methane released in many forested wetlands which have a large acrotelm, such as tropical and boreal peatlands, is very low, and these systems are net carbon sequesters. Although drainage of peatlands has been shown to reduce methane production, other studies have indicated that this may be more than compensated by the methane production in the associated drainage ditches. On the other hand, drainage of peatlands leads to rapid oxidation of the peat. Carbon dioxide release will increase dramatically to levels as high as 15 t C/ha/yr in the temperate zone, and 50 t C/ha/yr in the tropics through decomposition (Immirzi and Maltby, 1992) or more than 500t/ha through fire.

Peat as an energy source is only important for regional or domestic socio-economic reasons, because it is more expensive and emits more CO2 per unit energy than other fossil fuels. Peat has been used as an energy source for at least two millennia. At present peat only contributes marginally to worldwide energy production, but at the local and regional scale, it can still be an important energy source, particularly in Finland, Ireland, and Sweden. It also continues to be important in the Baltic States, Belarus and Russia. In recent years technical developments have led to lower, more competitive peat prices. As peat is more expensive and emits more CO_2 per unit energy than other fossil fuels, it is only of interest as an energy source for regional or domestic socio-economic reasons. In Finland and Ireland employment in rural areas is the most important motive for peat energy, whereas in Eastern European and Central Asian countries, independence from Russian oil and gas imports and the lack of foreign currency are important driving factors.

Peatlands provide many plant species that are utilized for food, fodder, construction and medicine. One of the oldest and most widespread uses of wild peatland plants is as straw and fodder for domestic animals. For example, in Poland 70% of the peatlands were used as hay meadows and pastures. A second important use, especially in the temperate and boreal zones of Eurasia, is the collection of wild edible berries and mushrooms. Cloudberries (Rubus chamaemorus) are an important dietary supplement for many Arctic residents as well as a source of cash income. In Finland, the estimated yield at the turn of the 20th century amounted to 90 million kg. In the north, wild plants are used for a great variety of purposes. Their use today is, however, declining, as is the knowledge required to find, identify and gather such plants.

Tropical peat swamp forests Tropical peat swamp forests provide a wide range of products, such as edible fruits, vegetables, medicinal and ritual plants, construction material (wood, rattans, bamboo), fibres and dyeing plants, firewood and traded products like rattans, timber and animals.

Important timber species are Ramin (*Gonostylus bancanus*) and Meranti (*Shorea* sp.). Both timber and non-timber forest products (NTFPs) are, besides providing employment and contributing to state and federal revenues, central to the well-being and livelihood of local indigenous communities, such as the Dayak and Iban. Socio-economic studies indicate that in Indonesia local communities may depend on the peat swamp forest for over 80% of their livelihoods, rather than depending on agriculture. NTFPs provide cash income to supplement daily expenses or are a 'safety net' in time of need. Moreover, they represent an essential part of subsistence, culture and heritage. To the Dayak, the forest landscape is not only viewed as a collection of biodiversity, but also as a meaningful object for their social, economic, politic, and cultural lives (the concept of "*Petak Ayungku*"). The forest functions are a strong chain that binds together all members of the Dayak communities of the past, present, and future (Colfer and Byron 2001).

3.4. INDIRECT USE OF PEAT LAND

Peatlands often form major components of local and regional hydrological systems and have the ability to purify water by removing pollutants (Joosten and Clarke 2002). Large peatland bodies may regulate the surface- and groundwater regime and mitigate droughts and floods. For example, tropical peat swamp forests serve as overflow areas in flooding periods, while in the dry season the stored water is slowly released (Klepper 1992). Riparian peatlands such as in the floodplain of the Pripyat River in Belarus store floodwaters, resulting in a downstream reduction of velocity and volume of peak discharges (Belakurov *et al.* 1998).

3.5. NON-USE VALUES OF PEAT LAND

The cultural and aesthetic values of natural and cultural peatlands offer high potential for ecotourism and recreation. The limited accessibility of mires and peatlands does not make them particularly suitable for mass recreation. Where facilities are available, however, large numbers of people may visit peatland reserves, e.g. the Everglades NP (USA), North York Moors NP (UK), and Spreewald Biosphere Reserve (Germany) (Joosten and Clarke 2002). In many other countries peatlands are an important part of the national park or protected area networks that attract tourists, such as in Canada, Finland, the Baltic countries, and the Netherlands.

Many more mires are used for low- intensity recreation. Such ecotourism can provide additional income, such as in the Tasek Bera Ramsar site in Peninsular Malaysia where local communities earn additional income by selling traditional handicrafts and guiding boat tours through the swamps (Santharamohana 2003). In Indonesia, Orangutan rehabilitation centres in some peat swamp forest reserves (e.g. Tanjung Putting, Central Kalimantan) attract local and international tourists. Much of the potential of peatlands as special and intriguing habitats remains unexplored, perhaps also because there is limited experience with the special facilities that could make them more accessible and attractive to visitors.

Relatively few people live entirely from and in peatlands. For many more people, peatlands provide for part of their livelihoods. In addition, they are part of their traditions and have a special place within the ancestral land area, being part of their spiritual and aesthetic world, frequently occurring in folklore, literature, paintings, and other art (Joosten and Clarke

2002). Peatlands are valuable for education and research, since they contain important archives of cultural and environmental history reaching back more than 10,000 years.

Fossils in the peat matrix, including pollen, plant remains, archaeological artefacts and even human sacrifices, reveal the ecological and cultural history of the peatland itself, its surroundings, and even more distant regions (Joosten and Clarke 2002).

The cultural and aesthetic values of natural and cultural peatlands offer high potential for ecotourism and recreation. The limited accessibility of mires and peatlands does not make them particularly suitable for mass recreation. Where facilities are available, however, large numbers of people may visit peatland reserves, e.g. the Everglades NP (USA), North York Moors NP (UK), and Spreewald Biosphere Reserve (Germany) (Joosten and Clarke 2002). In many other countries peatlands are an important part of the national park or protected area networks that attract tourists, such as in Canada, Finland, the Baltic countries, and the Netherlands.

Many more mires are used for low- intensity recreation. Such ecotourism can provide additional income, such as in the Tasek Bera Ramsar site in Peninsular Malaysia where local communities earn additional income by selling traditional handicrafts and guiding boat tours through the swamps (Santharamohana 2003).

In Indonesia, Orangutan rehabilitation centres in some peat swamp forest reserves (e.g. Tanjung Putting, Central Kalimantan) attract local and international tourists. Much of the potential of peatlands as special and intriguing habitats remains unexplored, perhaps also because there is limited experience with the special facilities that could make them more accessible and attractive to visitors.

CONCLUSIONS

Peatlands and people are connected by a long history of cultural development; the livelihoods of substantial parts of rural populations in both developed and developing economies still significantly depend on peatlands. From the tropics to the Arctic the environmental security and livelihoods of indigenous cultures and local communities depend on peatland ecosystem services and the steady supply of natural peatland resources. The value of peatlands as an ecosystem providing crucial ecological, hydrological and other services has generally been disregarded. People have commonly treated peatlands as wastelands, using them in many destructive ways, without taking the long-term environmental and related socio-economic impacts into account.

The main human impacts on peatlands include drainage for agriculture, cattle ranching and forestry, peat extraction, infrastructure developments, pollution and fires. Deterioration of peatlands has resulted in significant economic losses and social detriment, and has contributed to tensions between key stakeholders at local, regional and international levels. Lack of awareness and insufficient knowledge of peatland ecology and hydrology have been major root causes of peatland deterioration. The key economic, cultural and environmental role of peatlands in many human societies calls for a "wise use" approach that minimizes irreversible damage and sustains their capacity to deliver ecosystem services and resources for future generations.

REFERENCES

Anderson, J.A.R., 1963. The flora of the peat swamp forest of and Brunei including a catalogue of all recorded species of flowering plants, ferns and fern allies. *Gardens Bulletin* (Singapore), 29: 131-228.

Appanah, S. Ismail H., Samsudin, M. and Sadali, S. 1999. Flora survey in North Selangor Peat swamp forest. In *sustainable management of Peat swamp Forest in Malaysia.* Forest Department , Kuala Lumpur.

Basiron, Y. 2007. Palm oil production through sustainable plantations. *European Journal of Lipid Science and Technology,* 109: 289-295.

Belokurov, A., Innanen, S., Koc, A., Kordik, J., Szabo, T. Zalatnay, J. and Zellei, A. 1998. Framework for an integrated land-use plan for the Mid-Yaselda area in Belarus. *EPCEM,* Leiden.

Bermudez, A., Foco, G., Bottini, S. B. Infinite Dilution Activity Coefficients in Tributyl Phosphate and Triacetín, *J. Chem. Eng. Data,* 45(6),1105-1107 (2000).

Chai, PPK., Lee, BMH., Ismawi, O. 1989. Native medicinal plants of Sarawak. *Forest Department Sarawak.*

FAO. 1974. The Peat swamps of Sarawak and their potential for development. FAO Report No 3. *FAO Forestry and Forest Industries Development Project,* Malaysia. Kuala Lumpur

Giesen, W. 1991. Berbak Wildlife Reserve, Jambi Sumatra. *PHPA/AWB Sumatra wetlands project* Report No 13. AWB, Bogor.

Gorham, E. 1991. Northern Peatlands: Role in the carbon cycle and probable responses to climatic warming. *Ecological Applications,* 1: 182-195

Greenley, D. A., Walsh, R. G., & Young, R. A. 1981. Option value: empirical evidence from a case study of recreation and water quality. *The Quarterly Journal of Economics,* 96, 657–672.

Hofstede, R., Segaraa, P. and Mena, P.V. 2002. Los Páramos del Mundo. *Global Peatland* initiative/NCIUCN/ EcoCiencia, Quito.

Hooijer, A., Silvius, M., Wösten, H.D. and Page, S. 2006. PEAT-CO2, Assessment of CO2 emissions from drained peatlands in SE Asia. *Delft Hydraulics report* Q3943.

IGBP Terrestrial Carbon Working Group. 1998. The Terrestrial Carbon Cycle: Implications for the Kyoto Protocol. *Science* 280:1393-1394

Immirzi, C.P. and Maltby, E., 1992. The Global Status of Peatlands and their Role in the Carbon Cycle. Wetlands Ecosystems Research Group, Report 11, Department of Geography, University of Exeter, UK, 118 pp.

Joosten, H. and Clarke, D. 2002. Wise use of mires and peatlands: background and principles. International Mire Conservation Group/International Peat Society, Saarijärvi.

JRC 2003. GLC 2000 – Global land cover for the year 2000. European *Commission Joint Research Centre Publication* EUR 20849 EN).

Krutilla, J.V. 1967. Conservation reconsidered. *American Economic Review* 57 (3): 777–86.

Kumari, K. 1995. "An Environmental and Economic Assessment of Forest Management Options: A Case Study in

Malaysia". The World Bank: Environmental Department Papers No. 026. In: Woon et al., 1999.

Meijaard, E. 1995. The Importance of swamp forest for the conservation of the orang Utan (Pongo pygmaeus) in Kalimantan, Indonesia. In J.O. Rieley and S.E. Page (eds.) *Tropical Peatlands*. Samara Press.

Melling, L., 1999. Sustainable Agriculture Development on Peatland. Paper presented at the Workshop on Working Towards Integrated Peatland Management for Sustainable Development Kuching. 17-18 August 1999.

Neuzil, SG. 1995. Onset and rate of peat and carbon accumulation in four domed Ombrogenous peat deposits, Indonesia. In *J.O. Rieley and S.E. Page* (eds.) *Tropical Peatlands*. Samara Press.

Ng, PKL, Tay, JB, Lim, KKP and CM, Yang. 1992. The conservation of the fish and other aquatic fauna of the North Selangor Peat Swamp Forest and adjacent areas. AWB Kuala Lumpur

Patterson, J. 1999. Wetlands and Climate Change: Feasibility investigation of giving credit for conserving wetlands as carbon sinks. *Wetlands International,* Ottawa, Canada.

Prentice C. and Aikanathan S. (1989) *A preliminary faunal assessment of the North Selangor Peat swamp Forest.* Asian Wetland Bureau, Kuala Lumpur.

Rieley, J.O. and Page, S.E. (1997) Biodiversity and Sustainability of Tropical Peatlands. Proceedings of the International Symposium on Biodiversity, Environmental Importance and Sustainability of Tropical peat and Peatlands, Palangka Raya, Central Kalimantan, 4-8 September 1995. Samara Publishing Limited, Cardigan. UK.

Rieley, J.O., Ahmad-Shah, A. A. and Brady, M.A. 1996. The extent and nature of tropical peat swamps. In: *Tropical Lowland Peatlands of Southeast Asia.* Eds.: E. Maltby, C.P. Immirzi and R.J. Safford. July 1992. IUCN, Gland, Switzerland, pp. 17-53.

Santharamohana, M. 2003. Knowledge, Culture and Beliefs of the Semelai People of Tasek Bera. Wetlands International, Kuala Lumpur, Malaysia.

Shamsudin, I. 1995. Tree species in Peat Swamp Forest in Peninsular Malaysia. In J.O. Rieley and S.E. Page (eds.) *Tropical Peatlands.* Samara Press.

Succow, M. and H. Joosten (Eds.) 2001. *Landschaftsökologische Moorkunde.* 2nd Ed. Schweizerbart, Stuttgart.

Urapeepatanapong, C. 1996. *Flora of Narathiwat peat swamp forest.* Royal Thai Forest Department, Bangkok

Whitmore, TC. 1984 *Tropical Rainforests of the Far East.* Clarendon Press, Oxford.

Zakaria, I. 1999. Survey of fish fauna in peat swamp forests. In *sustainable management of peat swamp forest in Malaysia.* Forest Department, Kuala Lumpur

PART 2: TRAFICABILITY (MECHANICAL PROPERTIES AND CHEMICAL AND BIOLOGICAL CHARACTERISTICE)

Chapter 4

TRAFICABILITY OF PEAT

4.1. INTRODUCTION

Peat and organic soils commonly occur as extremely soft, wet, unconsolidated surface deposits that are integral parts of the wetland systems. These soils are found in many countries throughout the world. In the US peat is found in 42 states with a total acreage of 30 million hectares. Canada and Russia are the two countries with a large area of peat, 170 and 150 million hectares respectively (Hartlen and Wolski 1996). For the case of tropical peat, or tropical peat lands, the total world coverage is about 30 million hectares, two thirds of which are in Southeast Asia. Malaysia has some 3 million hectares - about 8% of the land area is covered with tropical peat. While in Indonesia peat covers about 26 million hectare of the country land area, with almost half of the peat land total is found in Indonesia's Kalimantan (Bujang, 2006). These soils are generally referred to as problematic soils due to their high compressibility and low shear strength. Access to these surface deposits are usually very difficult as the water table will be at, near or above the ground surface. The thickness of this deposit varies from just about 1 m to more than 20 m thick. Pressures on the land use by industry, vehicle mobility for mechanization, housing and infrastructure are leading to more frequent utilization of such marginal grounds. It is therefore necessary to expand the knowledge of their geotechnical properties and mechanical behavior, in particular those in relation to their deformation and shear strength characteristics, and subsequently devises suitable design parameters and construction techniques on these materials.

Traficability of soil is defined as the resistance of the soil which is mainly prevent the sinkage of the static or dynamic object of the soil. Peat swamp resistance depends on the thickness of the surface mat which is normally developed by the presence of the root of grass. Figure 4.1 shows the peat structure.

Peat occurs in both highlands and lowlands in Malaysia. The highlands peat is not extensive, whereas, the lowlands peat occurs almost entirely in low-lying area due to poorly drained depressions or basins in the coastal areas. The general peat terrain condition is typically very high in water content and usually in low bearing capacity. The highland peat traficability is good enough to provide the resistance to vehicle sinkage during transportation. On other hand, the lowland peat is very poor to provide the resistance to vehicle sinkage during transportation.

Based on the traficability of the peat swamp, the vehicle for doing the transportation operation or any other purpose should be fitted with tracks to produce low average ground pressure ranging from 9.81 to 24.5kN/m^2 for performing transportation activities.

In view of the diversity of the peat terrain, it seems logical to assume that no one vehicle will ever be developed which will be equally or even acceptably efficient everywhere - at least not within realistic economic constraints.

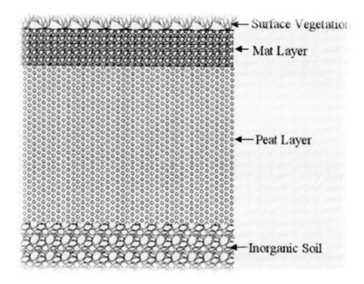

Figure 4.1. Structure of peat terrain.

The most successful vehicle in any one area, however, will be that which combines the requirements imposed by the various types of terrain encountered. It need not, then, be the one with the greatest mobility, but it will be the one with the best combination of mobility, speed, ride characteristics, serviceability, reliability, load carrying ability, and cost of service. The study of these interrelations is known as performance system analysis of the vehicle. For a specific vehicle, these qualities may sometimes be inferred by a study of design features such as: ground pressure, power-to-weight ratio, tracks or wheels, tracks or wheels contact area, front or rear drive, approach angle of the track, floatation capabilities, maximum and typical operating speeds, and transmission type. Figure 4.2 shows the highland peat. MacFarlane (1969) reported that the bearing capacity of highland peat range from 25 to 50 kN/m^2 and of lowland peat range from 10 to 25 kN/m^2. *In situ* the vane test result on highland peat showed that the shearing strength of the peat terrain in the ranged from 1.91 to 7.18kN/m^2 (Ataur et al.2004). On other hand, the lowland peat is very poor to provide the resistance to vehicle sinkage during transportation. As the peat land has high neutrients and good for plantation, there are 1.2million hectares of new oil palm plantations is expand on peat land of 600,000 hectares in Sabah and Sarawak and 400,000 hectares in peninsular Malaysia and 200,000 hectares conversion from rubber and coca plantations to achieve the production target of 16.66 million tons oil palm in 2010 (MPOB).

Figure 4.2. Highland peat.

Figure 4.3. Low land peat swamp.

4.2. DRAINAGE DEVELOPMENT AND USE OF CANALS

Disturbing the hydrological balance is one of the core threats to peatland ecosystems. When peat layers are exposed to oxygen, they start to decompose, dry out and become more susceptible to fires. This results in huge emissions of carbon dioxide. Excessive drainage can also cause a shrinkage or loss of wetland area, a reduction of water levels in adjacent wetlands and mineral soils and a decrease in water quality. Consequently, as peat subsides, the depth of the fertile topsoil also decreases. This means that further drainage, cultivation and pasture renewal are needed to maintain productivity, therefore increasing the cost to farmers (Wetlands International, 2007; Hooijer et al, 2006; ASEAN, 2006). Moreover, drainage increases the flood risk in wet periods, as the dry peat cannot take up the excess water.

Almost every form of development of peatland involves drainage of the peatland itself and/or its surrounding area to some extent (ASEAN, 2007). For agriculture and infrastructure projects, drainage has been practiced for many years. Traditionally, farmers used to develop relatively small, closed ended canals. Many more canals appeared and continue to appear for logging activities, both to drain the land as well as to transport the logs. However, the majority of canals in the area appeared as part of the MRP in mid nineties. In total, 4600 km of canals were dug in Central Kalimantan in order to drain the peatland and prepare it for agriculture (Wösten, 2005). Even though the MRP was a failure, the canals remain in place until today, creating a major drainage problem in the region. Nowadays, drainage canals still serve for agriculture, logging, fisheries and transportation both for people and goods (Page et al, 2002). Many of these canals are owned and operated by local villagers (CKPP web page). It is still a matter of debate whether some degree of drainage can be carried out which will avoid irreversible damage to the ecosystem. Meanwhile, canals still continue to be built, supported by the government as a result of conflicting objectives of economic development and conservation.

4.3. CHEMICAL AND BIOLOGICAL CHARACTERISTICS OF PEAT LAND

Tropical peatlands or peat swamp forests have developed primarily in the coastal lowland plains in-between major rivers (ASEAN, 2006). Technically, peat is partially decomposed organic matter that accumulates over thousands of years due to the lack of oxygen under waterlogged conditions. Tropical peatlands possess specific chemical characteristics. They are very acid and nutrient poor. Dry material of tropical peat consists of 50 to 60 percent carbon (Hooijer et al, 2006). Therefore, tropical peatlands store 2-6000 tons of carbon per hectare (t C/ha) compared to the average of 270 t C/ha on average in the world's forest ecosystems (ASEAN, 2007).

The peat soil in undisturbed circumstances consists of 80 to 90 percent water. The average water table depth in a natural peatland is near the soil surface. Water logging is a prerequisite for the creation and preservation of peat. These processes are highly sensitive to changes in hydrology and (micro-) climate (Wetlands International, 2007; Joosten, Clarke. 2002). Water, peat and specific vegetation are strongly interconnected. Therefore removing any one of these components or disturbing the balance between them may fundamentally change the nature of peatlands (Wetlands International, 2007). Badly degraded peatlands are virtually useless for most purposes and very difficult, expensive and time-consuming to rehabilitate. When dry, peat hardly absorbs any water and can be easily eroded by rainfall or enflamed. Nowadays, there are large areas of repeatedly burnt unproductive peatland covered by grass or ferns which once held productive forest and provided many useful services.

4.3. ACIDITY (pH) OF PEAT SOIL BY MICROBIAL TREATMENT

Peatland is an accumulation of organic matter resulting from incomplete decomposition of plant material. Peat soil has specific characteristics, for example, low pH (+3), high organic matter and cation exchange capacity and low base saturation (Koesnandar *et al.*,2005). These characteristics reduce the availability of nutrients especially K, Ca, and Mg that are bound in such way that it is difficult for them to be utilized by plants. Utilization of peat soil for food crops cultivation is subject to some constraints owing to its chemical, physical and biological properties especially when it has not undergone further decomposition. Nutrient availability in peat soil, especially nitrogen and phosphorus is related closely with the degree of decomposition. Nitrogen in a peat soil comes from organic compounds of plant decomposition, biological nitrogen fixation, water irrigation and inorganic fertilizer (Laegreid *et al.*, 1999). Although the content of nutrients in peat soil is high, these are not available to plants because they are bound in organic forms. The application of N, P, K and S fertilizer to peat soil with a high C/N ratio will be ineffective because these elements will be used by the microbes for decomposition process prior to being utilized by plants. Therefore, the degree of peat soil decomposition influences the availability of nutrients, physical characteristics and microbe dynamics of a peat soil.

Diana et al. (2006) were conducted a series of experiments has been performed to utilize a consortium of microbes to increase the pH of peat soil in an original peatland in Siantan Hulu Pontianak, Indonesia. The outline of their experiments: Oil palm empty fruit bunch fibre, mesocarp fibre and palm kernel cake from the oil palm industry were used as carbon sources for microbial fermentation. These carbon sources were layered on the peat and inoculated with designed microbial consortia. The layers were then mixed thoroughly with 10-15 cm of the top peat soil layer and then covered with plastic. Soil characteristics (base saturation, cation exchange capacity and C/N ratio) and pH were analyzed weekly for 6 weeks. They reported that the activity of the microbial consortium significantly increased pH of the peat soil, regardless of the carbon source used.

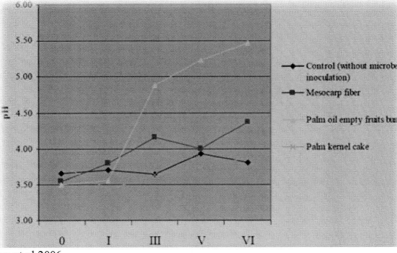

Source: Diana et al.2006.

Figure 4.4. Effect of carbon source utilization on pH of peat.

The highest increase in pH, from 3.50 to 5.47 after 6 weeks incubation, was obtained when oil palm empty fruit bunch fibre was used. The use of mesocarp fibre increased pH to 4.42 by the end of the third week and then it did not change up to the end of the experiment owing to the diminished fibre content of this source. Palm kernel cake could not be metabolized by microbes as shown by the fact that pH did not increase during incubation. In this case the pH of the peat soil was constant when either consortia or substrate were absent. Differences in the structure, characteristics and fibre content of the carbon sources used, as well as the organic materials in the peat soil, influenced the growth and metabolic pathways of the microbes (Fauzi *et al.*, 2006).

Table 4.1. The pH, Cation Exchange Capacity, Base Saturation and C/N ratio after incubation

Substrates	Final pH	Cation Exchange Capacity (me/100 g)	Base Saturation (%)	C/N Ratio
Control (without microbe inoculation)	3.80	79.46	20.67	58.67
Mesocarp Fibre	4.37	69.49	40.67	28.33
Oil Palm Empty Fruit Bunches	5.47	19.73	27	26
Palm Kernel Cake	3.87	86.07	20.33	31

Source: Diana et al. (2006).

A microbiological treatment of peat soil using cellulose-based waste of the palm oil industry as carbon source improved the peat soil characteristics, namely by increasing pH, decreasing cation exchange capacity and increasing base saturation and optimizing the C/N ratio. The results are summarized in Table 4.1 (Diana et al.2006). It appears that the microbial treatment on peat soil has advantages for the pH increment compared to physical treatment. In this regard, a highly efficient composting system utilizing oil palm mill fibre by solid state fermentation has been demonstrated using a thermophilic fungus *Chaetomium*sp.Nov.MS-017 (Suyanto *et al.*, 2003). The microbial approach to using oil palm mill fibre has been demonstrated as a potential tool for the production of compost with additional functions such as herbicide and anti plant pathogen, besides land fertilizing and soil conditioning properties (Mimura *et al.*, 2001).

4.3. MECHANICAL PROPERTIES OF PEAT SWAMP

Early research on peat strength indicates some confusion as to whether peat should be treated as a frictional material like sand or cohesive like clay. Commonly, surface peats are encountered as submerged surface deposits. Because of their low unit weight and submergence, such deposits develop very low vertical effective stresses for consolidation and the associated peat exhibit high porosities and hydraulic conductivities comparable to those of fine sand or silty sand (Dhowian and Edil 1980). Such a material can be expected to behave "drained" like sand when subjected to shear loading. However, with consolidation, porosity decreases rapidly and hydraulic conductivity becomes comparable to that of clay. There is a rapid transition immediately from a well drained material to an undrained material (Edil *et al.*,

1994). Determination of shear strength parameters for organic soils, as with other soils, is important and somehow a difficult job in geotechnical engineering.

The assessment of peat terrain features, an indirect approach must be used, whereas in other arried out by direct observation. Table 4.2 lists the terrain features and corresponding methods by which individual features can be detected. For organic soils, several methods have been used to determine the undrained shear strength in the laboratory namely Swedish fall-eone test, triaxial test, shear box test and vane shear test. For the case of field tests, field vane and Dutch Cone Penetration tests are often used.

Ataur et al. (2004) has been measured the mechanical properties of peat swamp as shown in Figure 4.5 in terms of the bulk density, sinkage stiffness, cohesiveness, internal friction angle and modulus of deformation. Field tests were carried out at Sepang swamp peat area, located about 45km from Kuala Lumpur, Malaysia.

Table 4.2. Methods of observation and evaluation of terrain features

Terrain feature	Method of analysis
Vegetal cover	Direct observation
Peat depth	Manual probing with rods
Peat structure	Radforth classification system
Shear strength	Shear vane and direct shear box test analysis
Cohesiveness	Direct shear box test
Cone index	Cone penetrometer
Peat water content	Sampling and drying or nuclear moisture meter
Sinkage parameters (surface mat and internal peat stiffness)	Plate sinkage test
Water acidity (pH) measurement	Microbiological treatment

Figure 4.5. Moderest peat swamp.

Determination the moisture content and bulk density, cohesiveness, internal friction angle and shear deformation modulus, vane shearing strength, surface mat stiffness and underlying stiffness of swamp peat was carried out both in situ and in laboratory. In-situ determination for vane shearing strength, surface mat stiffness, and underlying stiffness were carried out at 100mm, 250mm and 400mm depth in three replications at the center location between adjacent palm rows within each sub-block in the field for the un-drained and drained conditions. Similarly, undisturbed samples volume of $216cm^3$ were taken at the mentioned depths and points in the fields in three replications for the determination of moisture content, bulk density, cohesiveness, internal friction angle, shear deformation modulus of peat. The samples were wrapped with aluminum foil and sealed in a plastic container before were immediately taken to the laboratory for the relevant analysis.

4.3.1. Deformation Characteristics of Peat

Compression Index (C_c)

An effort is made to correlate compression index, c'' with liquid limit, wl' void ratio, e, and ratio of $c / (1 + e)$. Farrell *et al.* (1994) considered the empirical relationship between the compression index and the liquid limit suggested by Skempton and Petley (1970) for organic soils (equation 4.1). Hobbs (1986) estimated the compression index for fen peat using equation (4.2), which gave a slightly lower value of C_c, of tropical peat samples tested however were apparently a little higher than the above two relationships

$$C_e = 0.009(W_L - 10) \tag{4.1}$$

$$C_e = 0.007(W_L - 10) \tag{4.2}$$

The Cc values of the tropical peat studied range from 1 to 3, much higher than sedimentary soil such as clay whose c, is only 0.2 - 0.8. It is of interest to note that the c, of the Irish peat range from 1 - 4 which is quite close to the tropical peat (Farrel, 1994). Azzouz *et al.* (1976) reported the following relationship for organic soil and peat, where, W_n is soil natural water content in percent. Note that the natural water content of the peat studied range from 200 % - 800 %.

$$C_e = 0,0115(W_n) \tag{4.3}$$

Void Ratio with Liquid Limit and Natural Water Content

The initial void ratios versus liquid limits of the peat and organic soils from several sites in Malaysia together with the normally consolidated peat found by Miyakawa (1960), and Skempton and Petley (1970). Void ratio of the tropical peat studied is found to range from 1.5 - 6, that is for the case of amorphous peat. For the case of fibrous peat it can be as high as 25. Such high void ratios gives rise to phenomenally high water contents. For comparison, Malaysian marine clay for instance, has an initial void ratio in the range of 1.5 to 2.5. The natural void ratios of the peat indicate their higher compressibility. The empherical equation to determine the void ratio (Bujang, 2006):

$$e_o = \frac{30.65 \left(\omega_l + 0.88\right)^{0.116} - 30}{1.12} \tag{4.4}$$

4.3.2. Surface Mat Stiffness and Underlying Stiffness of Peat

The area was heavily infested with palm roots, low shrubs, grasses, and sedges. The field conditions were wet and the water table was found to be 0 to 120 mm below the surface level.

The dominant features of this site may be described as high water content and weak underlying peat that could easily be disturbed by vehicle movements.

The surface mat and the peat deposit thickness were not distinct by visual observation. The surface mat thickness was about 50 to 250 mm at the center. The underlying peat deposit thickness for the whole area was about 500 to 1000 mm. The water field capacity was almost at saturation level and walking on such a terrain condition was only possible with the use of a specially made wooden clog as shown in Figure 4.6. The sinkage of any heavy object is in peat swamp is computed by the following equation:

$$z = \frac{-\left(\dfrac{k_p D_h}{4 m_m}\right) \pm \sqrt{\left[\left(\dfrac{k_p D_h}{4 m_m}\right)^2 + \dfrac{D_h}{m_m} P_0\right]}}{2} \qquad (4.5)$$

with

$$D_h = \frac{4 BL}{2(L + B)}$$

where, P_0 is the normal pressure of the vehicle in kN/m^2 and z is the sinkage in m, m_m is the surface mat stiffness in kN/m^3, k_p is the underlying peat stiffness in kN/m^3, and D_h is the hydraulic diameter in m, B and L are the width and ground contact length of the track in m, respectively. The sinkage test was conducted by using the author designed and developed pressure sinkage plate device as shown in Figure 4.7. The test results were verified with Wong et al. (1989) results. The overall area was divided into 3 equal area blocks and each block was again divided into 3 equal sub-blocks. Each of the sub-blocks was considered for the travelling path of the vehicle. Based on the experiment on the Sepang peat area, it is mentioned that the surface mat thickness was found of 50mm. Normally, the surface mat thickness was in the range of 100 to 250mm. Therefore, the surface mat thickness was considered in the range of 100 to 250 mm for the simulation of the vehicle tractive performance.

A specially developed apparatus as shown in Figure 4.7 was used to determine the surface mat stiffness and underlying stiffness of peat. The apparatus comprises of a proof ring with a dial gauge for measuring the force, a shaft with diameter of 15.9 mm and with length of 450 mm for transferring load to sinkage plate and a rectangular plate with size of $150x50x5mm^3$ for measuring the sinkage.

Three sinkage plates having a diameter of 100 m and 150 m and a size of $150x50 \ mm^2$ were used to determine the stiffness of peat. This apparatus was setup to run at a proving ring constant of 0.2 kg/division. Prior to the actual stiffness test, the proof ring dial gauge was calibrated. During the operation of the apparatus for the stiffness test, the long shaft with sinkage plate diameter of 100 mm was pushed down at a constant speed of 25 mm/sec to sink the plate. Reading on the pushing load in kg and sinkage in cm was recorded for every 20 mm plate sinkage. The parameter characterizing the behavior of the surface mat that is represented by 'm' was determined from the slope of the curve of the Figure 4.8.

The same test procedure was repeated for sinkage plates with diameters of 100 mm and 150 mm and a size of 150x50 mm^2 on different sampling points, and for different drainage conditions.

Figure 4.6. Tested swamp peat site before drainage.

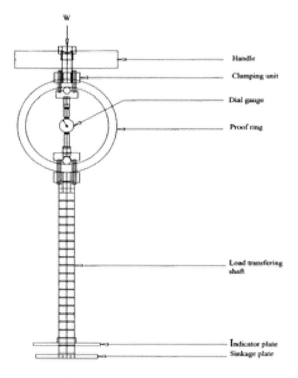

Figure 4.7. Bearing capacity measuring apparatus.

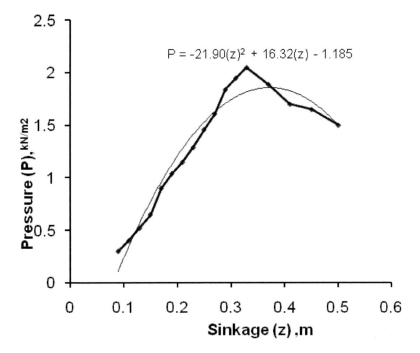

Figure 4.8. Typical load-sinkage trend of peat.

Surface mat stiffness and underlying stiffness of peat can be determined from the following equation of Wong (1982):

$$p = \frac{1}{100}\left[k_p z + mk_p z^2 (L/A)\right] \tag{4.6}$$

with

$$m_m = mk_p$$

where, p is the load applied on the handle of the apparatus in kN/m^2, A is the area of the plate in m^2, L is the perimeter of the plate in m, z is the sinkage in m, k_p is the stiffness of the peat in kN/m^3, m is a parameter characterizing the behavior of the mat, and m_m is surface mat stiffness in kN/m^3.

Equation (4.6) was used to predict the nominal ground pressure of the vehicle. The sinkage of the vehicle in the peat terrain can be determined by using the emphical equation of Rahman et al. (2004):

$$P = -21.9(z)^2 + 16.32(z) - 1.185 \tag{4.7}$$

where, P is the equivalent ground contact pressure of the vehicle in kN/m^2 and z is the sinkage in m.

Typical graph of load-sinkage of peat is shown in Figure 4.8. Surface mat stiffness of peat for un-drained and drained conditions at depths of 10, 25, and 40 cm are shown in Tables 4.3. Mean surface mat stiffness of peat increased from 27.07 to 44.51 kN/m^3 for plate with diameter of 10cm, 33.93 to 41.79 kN/m^3 for plate with diameter of 15 cm, and 32.54 to 50.57 kN/m^3 for plate with size of 15x5cm when field condition was changed from un-drained to drain.

Table 4.3. Variation of the surface mat stiffness of peat with depth

Sinkage (cm)	Plate size (cm)	Surface mat stiffness, m_m (kN/m^3) (kN/m^3)					
		Block 1		Block 2		Block 3	
		Before drainage	After Drainage	Before drainage	After drainage	Before drainage	After drainage
-10	D=10	31.2	52.46	40.21	59.49	34.01	49.4
	D=12	57	59.77	40.89	54.85	44	43
	LxB=15x5	41.4	66.76	39.54	97.68	43.88	68.5
-25	D=10	22.5	56.76	50.87	67.68	12.5	46.5
	D=12	37.67	39.21	53.91	54.86	13	33.4
	LxB=15x5	44.01	52.4	50.29	48.73	20.12	52.28
-40	D=10	15.76	23.8	26.44	30.43	10.12	14
	D=12	15.01	36.68	34.69	37.86	9.2	16.32
	LxB=15x5	15.71	24.85	24.01	27.57	13.90	16.4

Source: Ataur et al.(2004).

Table 4.4 shows that the main effects of sinkage, field condition and the interaction of sinkage*field condition*plate size were significant on peat surface mat stiffness.

The following observations were also made based on Table 4.3:

- Before drainage: Mean surface mat stiffness of peat decreased from 41.35 to 33.87 kN/m^3 and 33.87 to 18.31 kN/m^3 when the depth increased from 10 to 25 cm and 25 to 40 cm, respectively.

- After drainage: Mean surface mat stiffness of peat decreased from 61.32 to 50.2 kN/m^3 and 50.2 to 25.321 kN/m^3 when the depth increased from 10 to 25cm and 25 to 40 cm, respectively.

Table 4.4. ANOVA of the surface mat stiffness of peat

Source of variation	DF	SS	MS	F Value
Block	2	2491.50	1245.75	19.32**
Plate size	2	307.24	153.62	2.38
Sinkage	2	8192.68	4096.34	63.53**
Sinkage*Plate size	4	603.15	150.78	2.34
Field condition	1	2817.96	2817.96	43.70**
Field condition *Plate size	2	293.33	146.66	2.27
Sinkage* Field condition *Plate size	6	1202.09	200.34	3.11*
Error	34	2192.39	64.48	
Corrected Total	53	18100.30		

** Highly significant at probability level 1 % and *significant at probability level 5 %.

The significant effect of sinkage indicates that surface mat stiffness was decreased with increasing of sinkage. Result from $LSD_{(Pr<0.05)}$ shows that mean surface mat stiffness decreased from 51.336 to 42.043 kN/m^3 and 42.04 to 21.831 kN/m^3 when sinkage increased from 10 to 25 cm and 25 to 40cm, respectively. It indicates that within the depths from 10 to 25cm, the peat soil strength is provided mainly by the surface mat due to tension. Beyond this depth the strength was provided by the underlying weak and low bearing capacity peat.

Again result from $LSD_{(Pr<0.05)}$ shows that mean surface mat stiffness increased from 31.18 to 45.63 kN/m^3 when field condition was changed from un-drained to drained.

The significant effect of drainage indicates that the surface mat stiffness increased with draw-down of water table by drainage. Due to the drainage of the tested site the effective weight of the soil increased and caused consolidation of the fill and other soft underlying materials.

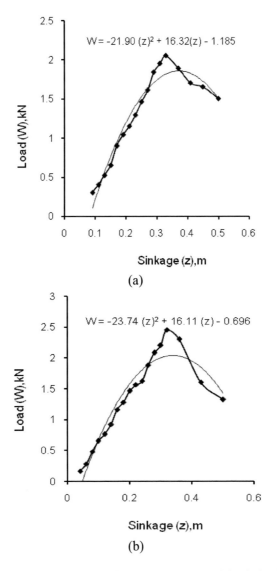

Figure 4.9. Typical load-sinkages trend of peat for (a) un-drained and (b) drained.

Underlying stiffness of peat for before and after drainaged conditions at depths of 10, 25, and 40 cm are shown in Table 4.5. Mean underlying stiffness of peat increased from 224.38 to 356.95kN/m^3 for plate with diameter of 10 cm, 274.15 to 378.85kN/m^3 for plate with diameter of 15cm, and 257.48 to 404.79kN/m^3 for plate with size of 15x5cm when field condition was changed from un-drained to drain.

The following observations were also made based on Table 4.5:

- Before drainage: Mean underlying stiffness of peat increased from 221.3 to 252.14kN/m^3 and 252.14 to 282.55kN/m^3 when the depth increased from 10 to 25cm and 25 to 40cm, respectively.
- After drainage: Mean underlying stiffness of peat increased from 356.76 to 370kN/m^3 and 370 to 413.79kN/m^3 when the depth increased from 10 to 25cm and 25 to 40cm, respectively.

Table 4.5. Variation of the underlying stiffness of peat with depth

Sinkage (cm)	Plate size (cm)	Peat stiffness, k_P (kN/m^3)					
		Block 1		Block 2		Block 3	
		Before drainage	After Drainage	Before drainage	After drainage	Before drainage	After drainage
-10	D=10	228.58	249.01	196.91	339.8	168.65	473.33
	D=12	216.55	342.36	235.82	222.1	212.96	374.66
	LxB=15x5	283.59	465.18	222.39	288.68	226.24	455.72
-25	D=10	260.68	186.68	218.04	348.21	176.43	523.16
	D=12	267.85	370.46	265.58	284.4	293.78	437.92
	LxB=15x5	255.94	387.90	277.54	301.82	253.43	489.85
-40	D=10	333.0	380.00	256.94	380.19	180.21	530.26
	D=12	325.00	407.81	288.50	318.3	361.35	453.62
	LxB=15x5	215.86	451.89	291.94	304.3	290.41	497.77

Table 4.6. ANOVA of the underlying stiffness of peat soil

Source of Variation	DF	SS	MS	F Value
Block	2	51559.692	5779.85	5.32**
Field condition	1	221857.79	221857.79	45.82**
Plate size	2	7822.01	3911.00	0.81
Sinkage	2	32170.59	16085.29	3.32*
Field condition *Plate size	2	14094.20	7047.09	1.46
Sinkage*Plate size	4	10314.38	2578.59	0.53
Sinkage*Drainage*Plate size	6	2271.01	378.50	0.08
Error	34	164643.83	4842.46	
Corrected Total	53	504733.51		

** Highly significant at probability level 1 % ; * Significant at probability level 5 %.

Table 4.6 shows that mean effects of sinkage and field conditions were significant (Pr<0.01) on internal peat stiffness. Result from LSD$_{(Pr<0.05)}$ shows that mean underlying stiffness of peat increased from 289.03 to 311.09kN/m^3 and 311.09 to 348.19kN/m^3 when sinkage increased from 10 to 25 cm and 25 to 40cm, respectively.

Again result from LSD$_{(Pr<0.05)}$ shows that mean underlying stiffness of peat increased from 244.59 to 380.2kN/m^3 when field condition was changed from un-drained to drained.

4.3.3. Moisture Content and Bulk Density

Water table level is artificially maintained high and the forest floor is flooded for most of the year in order to reduce risks of fire. The lowest water table was at -30 cm below the peat surface and the moisture of peat above the water table level was at the lowest 70% w/w (from the 80% water holding capacity).

The water table in the peat area was below the peat surface only from February to May (from the middle to the end of the dry season), and the lowest water table level was -30 cm below the peat surface as shown in Figure 4.10, reported by Duong et al. 2006. Peat above the water table had a moisture content of 70-82%, which is about 80-100% of its water holding capacity.

The collected samples that were brought to the laboratory were initially weighed for their wet mass before placing in an electric oven at 105°C temperature. After twenty-four hours, the samples were taken out from the oven and weighted for their dry mass.

The sample moisture content is defined as the percentage ratio between the differences in sample wet mass and dry mass in gram with the sample wet mass in gram.

The sample bulk density was computed by dividing the sample dry mass in gm to the sample known volume in cm^3. The bulk density measurement was conducted for determining the bearing capacity of the terrain for the dynamic condition of the moving heavy object (vehicle) can be represented by the equation of Terzaghi (1966):

$$q_{do} = \left[\left(1 - \frac{0.3B}{L}\right)cN'_c + \left(0.5 - \frac{0.1B}{L'}\right)BN'_\gamma \gamma_d\right]$$

(4.8)

where, qdo is the peat terrain bearing capacity during dynamic condition in kN/m2 , N'γ , N'c and N'q are referred to as Terzaghi's bearing capacity factors , B is width of the footprint, c is the cohesiveness in kN/m2 and L is the length of the track ground contact length in m.

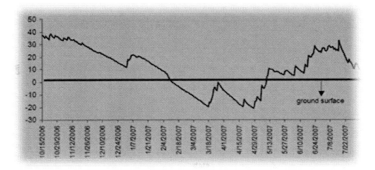

Source: Duong et al.2006.

Figure 4.10. Water table level in the research site in Vodoi National Park, Vietnam.

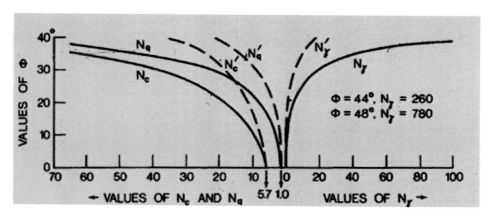

Figure 4.11. Variation of the Terzaghi bearing capacity factors with the angle of internal shearing resistance of soil (Source: Terzaghi, 1966).

The parameters N'_γ, N'_c and N'_q are referred to as Terzaghi's bearing capacity factors. The variation of N'_γ, N'_c and N'_q can be determined from the Figure 4.11, for the corresponding value of peat internal frictional angle φ.

Moisture content and bulk density of the peat for un-drained and drained conditions at depths of 10cm, 25cm, and 40cm are shown in Table 4.7. Examples of moisture content and bulk density variations with depth for peat under drained and un-drained field conditions are shown in Figures 4.12 – 4.13.

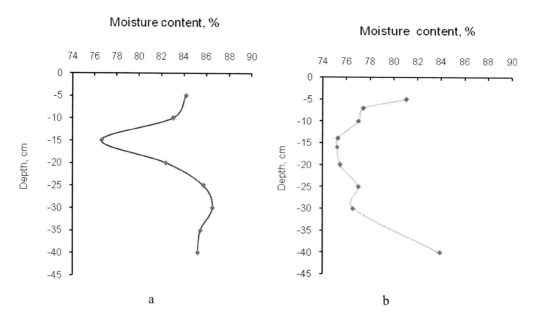

Figure 4.12. Typical trend of moisture content (a) un-drained and (b) drained.

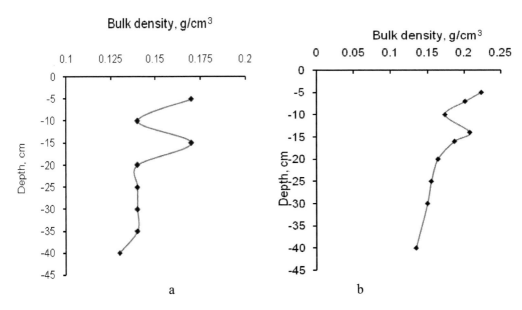

Figure 4.13. Typical trend of bulk density and depth (a) before drainage and (b) after drainage.

Mean moisture content decreased from 83.79% to 79.27% (5.4% decrement) and mean bulk density increased from 0.15g/cm^3 to 0.18g/cm^3 (22.62% increment) when the field condition was changed from un-drained to drain.

Table 4.7. Variation of moisture content and bulk density of peat

Location	Depth (cm)	Volume of the sample (cm^3)	Before drainage Moisture content (%)	Before drainage Bulk density (g/cm^3)	After drainage Moisture content (%)	After drainage Bulk density (g/cm^3)
1	-10	82.95	84.17	0.16	76.43	0.22
	-25	82.95	85.02	0.148	77.62	0.20
	-40	82.95	86.00	0.14	78.62	0.17
2	-10	82.95	82.40	0.17	75.37	0.20
	-25	82.95	85.76	0.14	77.66	0.18
	-40	82.95	86.53	0.14	79.55	0.16
3	-10	82.95	85.50	0.14	81.26	0.15
	-25	82.95	85.23	0.14	81.62	0.16
	-40	82.95	86.54	0.13	82.27	0.15

The following observations were also made based on Table 4.7:

- Before drainage: Mean moisture content of peat increased from 84.02 to 85.34% and 85.34 to 86.36% and mean bulk density of peat decreased from 0.16 to 0.14g/cm^3 and 0.14 to 0.13g/cm^3 when depth increased from 10 to 25cm and 25 to 40cm, respectively.
- After drainage: Mean moisture content of peat increased from 77.69 to 79% and 79 to 80.01% and bulk density of peat decreased from 0.19 to 0.18g/cm^3 and 0.18 to 0.16g/cm^3 when depth increased from 10 to 25cm and 25 to 40cm, respectively.

4.3.4. Natural Unit Weight

The natural unit weight depends upon the water and organic content of peat. Higher unit weights are usually associated with substantial inorganic content. Natural unit weights have been reported 0.04 g per cm^3 [Macfarlane, 1969]. The natural unit weight can be determined by the liquid displacement method. The dry unit weight can be determined by using the following equation of [Macferlane, 1969]:

$$\gamma_d = \frac{G}{1 + e}(\gamma_w)$$

(4.9)

where G is the specific gravity of soil solids, e is the void ratio, and γ_w is the unit weight of water. An experimental procedure for determining the specific gravity of peat solid was outline by Akroyed (1957). It was involved placing the peat sample in a flask, covering it with de-aired, filtered kerosene and applying a high vacuum until air bubbles cease to be emitted from the sample. The container was then filled with kerosene and permitted to reach a constant temperature. He reported that the specific gravity of the solids in peat grater than 1 (Akroyd 1957). The specific gravity of soil solids can be determined by using the equation of Akroyd (1957)):

$$G = \frac{weight\ \ of\ dry\ soil}{weight\ \ of\ \ker osene\ \ displaced} \times specific\ \ gravity\ \ of\ \ker osene$$

(4.10)

4.4. Cohesiveness, Internal Friction Angle and Shear Deformation Modulus

A Wykeham Farrance 25402 shear box apparatus shown in Figure 4.14 was used to determine cohesiveness, internal friction angle, and bulk deformation modulus of peat. The apparatus comprises of two equal halves open rectangular box for placing the test sample, a horizontal proof ring with a dial gauge for measuring shear displacement, a vertical proof ring with a dial gauge for measuring normal load, a hanger with dead weights for imposing normal load on the test sample, and a control-display box to control the overall test setup and display the imposed shear rate on the test sample. The apparatus was setup to run at a shear rate of 0.25 mm/sec and a maximum shear displacement of 12mm. Prior to the actual shear test, the prepared test sample in the rectangular box was subjected to a consolidation load of 1.0 kg for 24.0 hours. During the operation of the apparatus for the shear test, the top half of the rectangular box was kept stationery and the lower half was moved to shear the test sample. Readings on the shear displacement in cm and shearing strength in kN/m^2 was recorded for every minute until failure on the sample. The same test procedure was repeated for the consolidation load of 1.5 kg and 2.0 kg, and again repeated on samples taken from different depths, different sampling points, and different drainage conditions.

The shear deformation modulus of peat was determined by using the equation proposed by Wong et al., (1979):

$$K = -\frac{\sum\left(1-\frac{\tau}{\tau_{max}}\right)^2 j^2}{\sum\left(1-\frac{\tau}{\tau_{max}}\right)^2 j\left[\ln\left(1-\frac{\tau}{\tau_{max}}\right)\right]} \quad (4.11)$$

where, K is the shear deformation modulus in cm, τ_{max} is the maximum shear strength in kg/cm^2, τ is the measured shear stress in kg/cm^2 and j is the corresponding shear displacement in cm, respectively. The relationship between normal strength and shearing strength of peat for different samples was drawn as shown in Figure 4.14. From the interpretation of normal strength and shearing strength, the cohesiveness and the internal friction angle of peat were computed.

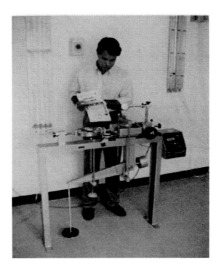

Figure 4.14. Wykeham Farrance 25402 shear box apparatus.

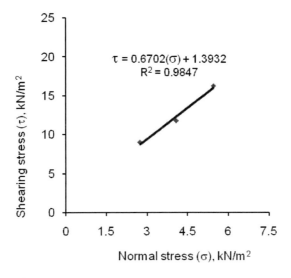

Figure 4.15. Typical trend of shearing stress versus normal stress.

Shearing strength of peat for un-drained and drained conditions at depths of 10 cm, 25 cm, and 40 cm and at normal loads of 1kg, 1.5kg, and 2kg are shown in Table 4.6.

Typical graph of consolidation and shearing strength versus normal load is shown in Figures 4.14 – 4.15. Mean shearing strength of peat increased from 14.03 to $16.22kN/m^2$ when the field condition was changed from un-drained to drain.

The following observation were made based on Table 4.8:

- Before drainage: Shearing strength of the peat soil increased from 10.68 to $13.63kN/m^2$ (decrement 27.62%) and 10.68 to $17.79kN/m^2$ (decrement 65.54%) when the normal stress increased from 2.725 to $4.08kN/m^2$ and 2.725 to $5.45kN/m^2$, respectively.
- After drainage: Shearing strength of the peat soil increased from 12.8 to $15.34kN/m^2$ (decrement 19.8%) and 12.8 to $20.48kN/m^2$ (decrement 60.0%) when the normal stress increased from 2.725 to 4.08 kN/m^2 and 2.725 to $5.45kN/m^2$, respectively.

Table 4.9 shows that the mean effect of field condition, depth, normal stress, and the interaction of depth and field condition on shearing strength of peat are highly significant. The significant (Pr<0.01) effect of field condition, depth, and normal stress indicate that the shearing strength of the peat soil was increased and thus may be due to the draw down of the water table by drainage.

The draw down of the water table resulted in increased consolidation of the fill and other soft underlying materials. Result from $LSD_{(Pr<0.05)}$ shows that the mean value of the shearing strength increased from 15.15 to $16.13kN/m^2$ and decreased from 15.15 to $13.907kN/m^2$ when the depth increased from 10 to 25cm and 10 to 40cm, respectively. Result from $LSD_{(Pr<0.05)}$ shows that the mean value of the shear strength increased 11.14 to $14.34kN/m^2$ and 14.34 to $19.14kN/m^2$ when the normal load increased from 1 to 1.5kg and 1.5 to 2kg, respectively.

Cohesiveness, internal friction angle, and shear deformation modulus of peat for un-drained and drained conditions at depths of below 10, 25, and 40 cm are shown in Table 4.10. Mean cohesiveness increased from 2.65 to $2.89kN/m^2$, mean internal friction angle increased from shear 22.33 to 23.76°, and mean shear deformation modulus decreased from 1.16 to 1.14cm when the field condition was changed from un-drained to drain.

The following observations were also made based on Table 4.10:

- *Before drainage:* Mean cohesiveness of peat decreased from 3.37 to $2.75kN/m^2$ and 2.75 to $1.77kN/m^2$, mean internal friction angle of peat decreased from 26.16 to 23.78° and 23.78 to 18.71°, and mean shear deformation modulus of peat increased from 1.12 to 1.17cm and 1.17 to 1.19cm when the depth increased from 10 to 25cm and 25 to 40cm, respectively.
- *After drainage:* Mean cohesiveness of peat decreased from 3.59 to $3.13kN/m^2$ and 3.23 to $1.95kN/m^2$, mean internal friction angle of peat decreased from 28.43 to 25.11° and 23.78 to 19.39°, and mean shear deformation modulus of peat increased from 1.1 to 1.14cm and 1.14 to 1.18cm when the depth increased from 10 to 25cm and 25 to 40cm, respectively.

Table 4.8. Variation of the shearing strength of peat in laboratory analysis

| Depth (cm) | Normal stress (kN/m^2) | Shearing strength (kN/m^2) | | | | | | | | Increase (%) |
| | | Block 1 | | Block 2 | | Block 3 | | Mean value | | |
		Before drainage	After drainage	Before drainage	After drainage	Before drainag	After drainage	Before drainage	After drainage	
-10	2.72	10.32	11.41	13.7	15.1	9.66	11.74	11.23	12.75	13.53
	4.08	11.30	13.01	17.7	18.05	12.04	13.23	13.68	14.76	7.89
	5.45	15.14	17.27	23.47	25.0	16.17	18.54	18.26	20.27	11.00
-25	2.72	10.89	12.0	13.14	13.75	11.04	14.1	11.69	13.28	13.60
	4.08	17.15	17.68	16.44	16.28	13.2	15.69	15.59	16.55	6.15
	5.45	19.01	21.2	20.18	22.85	18.3	20.21	19.16	21.42	11.79
-40	2.72	9.0	12.2	9.39	11.5	9.00	13.4	9.13	12.37	35.48
	4.08	11.79	16.4	11.79	13.0	11.23	14.9	11.62	14.77	27.10
	5.45	16.26	21.08	18.63	21.15	13.0	17.10	15.96	19.78	23.93

Table 4.9. ANOVA of the shearing strength of peat

Source of variation	DF	SS	MS	F Value
Block	2	74.28	37.14	20.92**
Depth	2	44.718	22.36	12.60**
Field condition	1	57.62	57.62	32.46**
Normal stress	2	510.11	255.05	143.70**
Depth* Field condition	2	11.80	5.90	3.32*
Depth*Normal stress	4	1.46	0.37	0.21
Depth*Normal stress* Field condition	4	1.46	0.36	0.21
Error	30	53.25	1.77	
Corrected Total	53	821.71		

** Highly significant at probability level 1% , and * Significant at probability level 5%.

Table 4.10. Variation of the cohesiveness, internal friction and shear deformation modulus

Depth (cm)	Block -1		Block-2		Block-3		Shear deformation modulus (cm)
	Cohesiveness (kN/m²)	Internal friction angle (degree)	Cohesiveness (kN/m²)	Internal friction angle (degree)	Cohesiveness (kN/m²)	Internal friction angle (degree)	
-10	5.06	27.22	.37	23.05	4.68	28.22	1.46
-25	3.33	17.66	2.64	21.5	3.48	24.18	1.31
-40	1.57	18.45	1.36	18.08	2.38	19.6	1.26

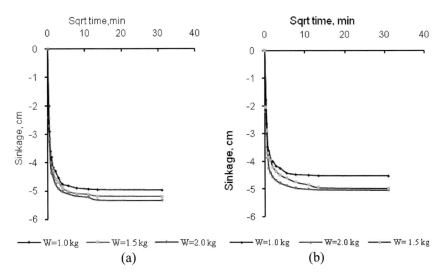

Figure 4.16. Typical consolidation test of peat (a) undrained; (b) drained.

4.5. *In-Situ* Shearing Strength

A RMU I012 digital vane test apparatus shown in Figure 4.18 was used to determine the *in-situ* shearing strength of peat. Figure 4.19 shows the scenerio of the vane test. The apparatus comprises a set of hollow rods for holding the vane blade, a twisting instrument on the upper end for twisting the rod set, and a digital measuring gearbox for measuring the twisting torque. Three vane blades having a diameter of 44 mm, 54mm, and 64mm were used to shear the peat. This apparatus was setup to run at a twisting torque ranging from 500 to 600 kg-cm at an accuracy of 1kg-cm resolution. Prior to the actual test, the digital measuring gearbox was calibrated.

Figure 4.18. RMU I012 digital vane shear test apparatus.

Figure 4.19. *In situ* shearing test determinng scenerio.

During the operation of the apparatus for the shear test, the twisting instrument was rotated at a speed of 12deg/sec to shear the peat. Reading on the twisting torque was recorded for every 7.5 seconds until a complete 360° rotation. The maximum twisting torque of a complete 360° rotation was recorded. The measured value was later used to compute the shear strength of the peat. The same test procedure was repeated for different depths, different blades with diameter of 54 mm and 64 mm, different sampling points, and different drainage conditions.

The shearing strength of the peat *in-situ* can be calculated using the following equation of MacFarlane (1969):

$$\tau_f = \frac{3T}{2\pi r^2 (2r+3h)} \qquad (4.12)$$

where, τ_f is the shear strength in kN/m^2, T is the torque in kN-m, r is the radius of the shear vane m, and h is the height of the shear vane in m.

Shearing strength of peat for un-drained and drained conditions at depths of 10, 25, and 40 cm are shown in Table 4.11. Mean shearing strength of peat increased from 1.96 to 2.29kN/m2 when the field condition was changed from un-drained to drain. Furthermore, mean shearing strength of peat decreased from 2.86 to 2.167kN/m2 (decrement 24.23%) and 2.167 to 1.78kN/m2 (decrement 17.85%) when vane blade size was increased from 4.4 to 5.4 cm and 5.4 to 6.4cm, respectively.

Table 4.11. Variation of the *in-situ* shear shearing strength of peat with depth

Depth (cm)	Vane blade size cm	Shearing strength (kN/m^2)					
		Block 1		Block 2		Block 3	
		Before drainage	After drainage	Before drainage	After drainage	Before drainage	After drainage
-10	D=4.4	1.99	4.08	3.77	2.83	3.77	4.48
	D=5.4	1.53	2.21	2.38	3.23	2.72	3.91
	D=6.4	1.53	1.74	1.94	2.14	2.04	2.24
-25	D=4.4	1.88	2.38	2.09	2.18	1.86	2.98
	D=5.4	1.98	2.30	1.80	2.12	2.52	2.86
	D=6.4	1.34	1.52	1.49	2.10	1.78	2.186
-40	D=4.4	1.57	2.59	1.77	1.98	1.57	1.84
	D=5.4	1.41	1.63	1.45	1.51	1.80	2.21
	D=6.4	1.73	1.48	1.61	1.68	1.67	2.14

The following observations were also made based on Table 4.11:

- Before drainage: Shearing strength of the peat soil increased from 2.41 to 1.86kN/m^2 and 1.86 to 1.62kN/m^2 when the depth increased from 10 to 25cm and 25 to 40cm, respectively.
- After drainage: Shearing strength of the peat soil increased from 2.98 to 2.29kN/m^2 and 2.29 to 1.89kN/m^2 when the depth increased from 10 to 25cm and 25 to 40cm, respectively.

Typical trend of *in-situ* shearing strength variations with depth for peat under drained and un-drained field conditions are shown in Figure 4.20.

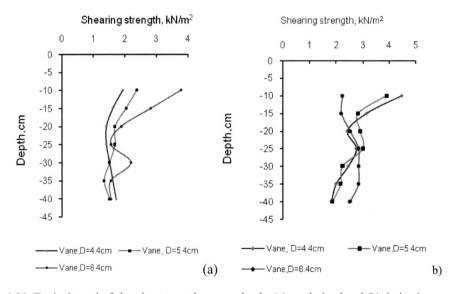

Figure 4.20. Typical trend of shearing strength versus depth, (a) un-drained and (b) drained.

Table 4.12 shows the mean effects of field conditions, depth, and blade size were highly significant (Pr <0.01) on *in-situ* shearing strength of peat. Result from LSD (Pr < 0.05) shows that mean shearing strength increased from 1.96 to 2.58 kN/m^2 when the field condition was changed from un-drained to drained. It may be due to the increase of consolidation rate. Result from LSD (Pr < 0.05) shows that mean *in-situ* shearing strength of peat decreased from 2.90 to 2.13 kN/m^2 and 2.13 to 1.78 kN/m^2 when the depth increased from–10 to –25 cm and -25 to –40 cm, respectively. It may be due to the draw down of the water table.

Table 4.12. ANOVA of the *in-situ* shear testing shearing strength of peat

Source of variation	DF	SS	MS	F Value
Block	2	7.85	3.92	7.20[*]
Field condition	1	5.23	5.23	9.60**
Depth	2	11.70	5.85	10.74**
Vane blade size	2	10.82	5.41	9.93**
Depth* Field condition	2	1.09	0.55	1.00
Depth*Plate size	4	6.64	1.66	3.05*
Depth*Plate size* Field condition	6	3.51	0.58	1.07
Error	30	16.34	0.54	
Corrected Total	53	66.70		

** Highly significant at probability level 1% and * significant at probability level 5%

Table 4.13. Peat terrain parameters

Parameters	Un-drained		Drained	
	Mean value	SD	Mean value	SD
ω, (%)	83.51	-	79.58	-
γ, (g/cm^3)	0.156	0.06	0.186	0.08
c, kN/m^2)	1.36	0.21	2.73	0.39
φ, (degree)	23.78	4.56	27.22	2.19
K_w, (cm)	1.19	0.10	1.12	0.17
m_m,(kN/m^3)	27.07	13.47	41.79	13.37
k_p, (kN/m^3)	224.38	52.84	356.8	74.27

CONCLUSION OF THIS CHAPTER

The mechanical properties of Sepang peat terrain were determined for evaluating the terrain traficability and vehicle mobility. The mechanical properties of Sepang peat terrain were determined by applying different techniques. Based on the mechanical properties the following conclusions were drawn:

1. Based on the results of mechanical properties of peat in the area studied, the mean values for moisture content, ω, and bulk density,γ, cohesiveness, c, internal friction angle,φ, and shear deformation modulus, K_W, and surface mat stiffness, and

underlying peat stiffness, k_p for un-drained and drained under the worst conditions are as stated in Table 4.13.

2. Bearing capacity for Sepang peat terrain was changed significantly from un-drained to drained terrain conditions.
3. Mechanical properties of the swamp peat terrain can be measured by developing an intelligent portable sinkage and strength testing machine.

REFERENCES

ASEAN (2006). *Rehabilitation and sustainable use of peatlands in South East Asia.* Full Project Brief for International Fund for Agricultural Development and Global Environment Facility. Appendix J: Summary of Country Components.

ASEAN (2007). *Rehabilitation and sustainable use of peatlands in South East Asia.* Full Project Brief for International Fund for Agricultural Development and Global Environment Facility.

Azzouz, A.S., KRIZEK, RJ. and RB. COROTIS. 1976. Regression analysis of soil compressibility. *Soils and Foundation* 16(2): 19-29.

Bujang B. K. Huat. 2006. Deformation and Shear Strength Characteristics of Some Tropical Peat and Organic Soils. Pertanika J. Sci. and Techno!. 14(1 and 2): 61 - 74 (2006).

Diana Nurani, Sih Parmiyatni, Heru Purwanta, Gatyo Angkoso, Koesnandar. 2006. Increase in ph of peat soil by microbial treatment.

Duong Minh Vien, Nguyen Minh Phuong, Jyrki Jauhiainen and Vo Thi Guong. 2006. *Carbon dioxide emission from peat land in relation to hydrology, peat moisture, humification at the Vodoi National Park*, Vietnam.

Farrell, E.R, C. O'Neill and Morris, A. 1994. Changes in the mechanical properties of soils with variation in organic content. In *Advances in Understanding and Modeling the Mechanical Behaviour of Peat*, p. 19-25. Balkema Rotterdam.

Fauzi Y., Widyastuti, Y.E., Satyawibawa, I. and Hartono, R. (2006). *Cultivation, product utilization and market analysis of oil palm.* Penebar Swadaya, Jakarta. (In Indonesian).

Hartlen, J and Wolski, J. 1996. *Embankments on Organic Soils.* Elsevier.

Hobbs, N. B. 1986. Morphology and the properties and behaviour of some British and foreign peats. *Quaterly Journal of Engineering Geology* 19: 7-80.

Hooijer, A., Silvius, M., Wösten, H. and Page, S. (2006). *PEAT-CO2, Assessment of CO2 emissions from drained peatlands in SE Asia.* Delft Hydraulics report.

Joosten, H. and Clarke, D. (2002). *Wise Use of Mires and Peatlands – Background and Principles including a Framework for Decision-Making.* International Mire Conservation Group and International Peat Society (eds). Greifswald, Germany: International Mire Conservation.

Koesnandar, Parmiyatni, S., Nurani, D., Wahyono, E. (2006). Government Role on Research and Application of Technology for Peatland Utilization. *National Seminar on peatlands and their problems.* March 21st 2006. University of Tanjungpura, Pontianak.(In Indonesian).

Laegreid, M., Bockman, O.C. and Kaarstad, O. (1999). *Agriculture, Fertilizers and the Environment.* Norsk Hydro ASA: CABI Publishing.

MacFarlane, I. C. 1969. *Muskeg Engineering Handbook.* University of Toronto Press,Toronto.

MPOB.2003.http://161.142.157.2/home2/home/stac03_prod1.htm. Date *of access*: January 2005.

Page, S.E., Rieley, J.O. (1998). *Tropical Peatlands:* A Review of Their Natural ResourceFunctions, with Particular Reference to Southeast Asia. *International Peat Journal, 8,* 95-106.

Skempton,A.W. and Petlev, D.J. 1970. Ignition loss and other properties of peats and clays from Avonmouth, King's Lynn and Cranberry Moss. *Geotechniques* 20(4): 34~356.

Suyanto, Ohtsuki, T., Ichiyazaki, S., Sadaharu, Subroto, A., Koesnandar and Mimura, A. (2003). Possibility of efficient composting on palm oil mill fibre by optimized solid state fermentation using thermophilic fungus *Chaetomium* sp., *Journal of Microbiology Indonesia,* Vol. 8 No. 2:57-62.

Wetlands International (2007). *Assessment on Peatlands, Biodiversity and Climate Change.*

Wong, J. Y. J., Radforth, R., and Preston-Thomas, J. 1982. Some further studies on the mechanical properties of muskeg in relation to vehicle mobility. *Journal of Terramechanics,* 19(2), pp.107-127.

Wong, J. Y., Garber M., Radforth, J.R and Dowell, J.T.1979. Characterization of the mechanical properties of muskeg with special reference to vehicle mobility. *Journal of Terramechnics,* 16(4): pp.163-180.

Wong, M.H. 1989. Workshop on tropical peat ecosystem in the costal areas of Peninsular Malaysia and southern Thailand. Malaysian Agricultural RandD Institute, 10th August.

PART 3:
MECHANIZATION: WHEELED VEHICLE, TRACKED VEHICLE AND INTELLIGENT AIR-CUSHION TRACKED VEHICLE

Chapter 5

MECHANIZATION OF PEAT SWAMP

5.1. INTRODUCTION

Transportation operation is an important problem on off-road vehicles in agriculture over the swamp peat terrain and is considered as one of the biggest issue in the many parts of the world. In view of the diversity of the peat terrain, it seems logical to assume that no one vehicle will ever be developed which will be equally or even acceptably efficient everywhere - at least not within realistic economic constraints. The most successful vehicle in any one area, however, will be that which combines the requirements imposed by the various types of terrain encountered. It need not, then, be the one with the greatest mobility, but it will be the one with the best combination of mobility, speed, ride characteristics, serviceability, reliability, load carrying ability, and cost of service. The study of these interrelations is known as performance system analysis of the vehicle. For a specific vehicle, these qualities may sometimes be inferred by a study of design features such as: ground pressure, power-to-weight ratio, tracks or wheels, tracks or wheels contact area, front or rear drive, approach angle of the track, floatation capabilities, maximum and typical operating speeds, and transmission type.

Designing and developing of a tracked vehicle requires a comprehensive understanding of the mechanical properties of the peat terrain under loading conditions similar to those imposed by the vehicle. As far as the prediction of vehicle tractive performance and mobility is concerned, the peat terrain response was characterized by mechanical properties such as shearing strength parameters and pressure-sinkage parameters. This study starts with the determination on the mechanical properties. The details of the mechanical properties has discussed in Chapter III. The measured mechanical properties are later used in the developed mathematical models to predict the tractive performance and engine power requirement for a prototype vehicle. Furthermore, the mathematical models are used to optimize the design and operating parameters of the prototype vehicle for both straight and turning motions. The established mathematical models may initiate other designers and manufacturers in optimizing and identifying the design parameters of any vehicle for low bearing capacity peat terrain. The established mathematical models may also initiate designers and manufactures in selecting the optimum turning radius and turning moment for maintaining the vehicle steady state turning and increasing the vehicle turning maneuverability. The following vehicles are considered for the collection-transportation of industrial and agriculture products on peat swamp:

1. Lower ground pressure (LGP) wheeled vehicle for highland peat terrain;
2. Tracked vehicle for moderate swamp peat;
3. Intelligent air cushion full tracked vehicle for peat swamp;
4. Semi wheeled tracked vehicle for moderate swamp peat;
5. Others peat vehicle.

All of these vehicles have been used to mechanize the palm oil, pine apple, sago and others on highland peat and peat swamp land.

Chapter 6

WHEEDED VEHICLE FOR HIGHLAND PEAT

6.1. INTRODUCTION

Typical peat characteristics as found in Malaysia are the presence of submerged and un-decomposed woods, stumps and logs. These submerged and unrecompensed logs or stumps impede the movement of machinery in the field. Other important characteristics are the very high ground water table, low bulk density and bearing capacity. Under the loaded surface some of peat soil may be at rest while others may move down. It is very difficult to manage any vehicle operation on peat terrain to do the transportation of palm oil fresh fruit bunches and other goods. This study investigates the LGP-30 wheeled vehicle tractive performance investigates through theoretically and experimentally. The major purpose of this study is to justify the LGP-30 wheeled vehicle suitability over the moderate peat terrain in Malaysia.

6.2. TRACTION MECHANICS

The motion of the vehicle's rolling wheel relative to the terrain is determined by analyzing the kinematics of the wheel. While, the tangential force (tractive force) is determined based on the tire-terrain interaction mechanism which could be achieved by reducing the inflation pressure of the tires. On a peat terrain, the performance of the vehicle is, to a great extent, dependent upon the manner in which the vehicle interacts with the terrain. The following assumptions are made to validate the mathematical model for wheeled vehicle:

- Based on the study of peat terrain mechanical properties addressed in ref. [7], the critical slip sinkage and ground contact pressure are considered to be 120 mm and 12 kN/m^2, respectively.
- Terrain reaction at all points on the contact patch is purely radial and is equal to the normal pressure beneath a horizontal contact area of the vehicle.
- The perimeter of the tire contact area and the velocity of the vehicle are considered to be constant.
- The rotational inertia of the vehicle rotating parts is neglected as it has no significant effect on the vehicle performance for straight motion.

The vehicle's mobility is limited by the terrain capacity. In general, if the vehicle ground contact pressure is more than the normal ground pressure of the terrain, the vehicle is at risk to operate on the low bearing capacity peat terrain. In order to increase the floatation capacity of the vehicle and decrease the vehicle ground contact pressure on the tire-terrain interface, the tire inflation pressure is assumed to be reduced. Therefore, a portion of the circumference of the tire will be flattened. The total pressure of the tire, P_0 on the peat terrain will be the sum of the inflation pressure P_i and the pressure due to carcass stiffness P_c. Based on the characteristics of low bearing capacity peat terrain, the vehicle's tracked system would not potentially be able to traverse on the peat terrain if the vehicle ground contact pressure for the wheeled system P_{cw} of is greater than the normal ground pressure of the terrain $P_{gt.}$. Therefore, the wheel system will be potential if

$$P_i + P_c = P_{g_t} \leq P_{vc} \qquad (6.1)$$

where

$$P_{g_t} = (k_p z_o + \frac{4}{D_h} m_m z_0^2)$$

$$D_h = \frac{BH}{(H + B)} \cdot P_{gt} \qquad \text{and}$$

is the normal ground pressure in kN/m^2 and z_o the critical sinkage (i.e, rut of depth or surface mat thickness) in m, m_m the surface mat stiffness in kN/m^3, k_p the underlying peat stiffness in kN/m^3, and D_h the hydraulic diameter in m, while B and H are the width and ground contact length of the wheel in m, respectively. *The critical sinkage of the vehicle in this study is defined as the surface mat thickness of the peat terrain which is considered the allowable sinakge of the vehicle.* It is anticipated that if the critical sinkage of the vehicle is more than the allowable sinkage, the vehicle will be highly risk to traverse and it would be immobilized as the slippage increases.

Slippage

Slip is one of the functional parameters for the vehicle traction mechanism. As the tractive force developed by a tire is proportional to the applied wheel torque under steady state conditions. It is defined as if the vehicle remains stuck and wheel slip continues, the driving force is reduced drastically.

$$i_{rd} = \left(1 - \frac{V}{R_o \omega}\right) \times 100\ \%$$

$$(6.2)$$

The slip of the wheel could determine by using the cycloid principle. *A cycloid is the curve defined by a fixed point on a rim of the wheel as it rolls, or, more precisely, the locus of a point on the rim of a circle rolling along a straight line. Points of rolling rim describe a cycloid.* Consider a wheel of radius R_0 which is free to roll along the x-axis. As the wheel turns, a point P on the tire traces out a curve.

Assume P is initially at the origin and let C and T are located as indicated in Figure 6.1, with ϕ denoting the radian measure of angle TCP. Then the arc PT and the segment OT have the same length such that the center C of the rolling circle is at (R_0ϕ, R_0).

Using trigonometry, it could be concluded that

$$E_{evaporation} = f(T) \qquad (6.3)$$

$$Q_f = V \times A_{drain} \qquad (6.4)$$

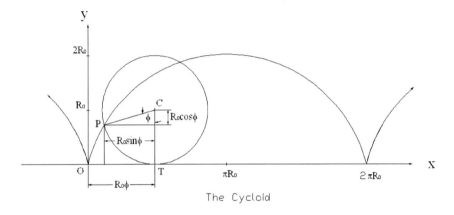

Figure 6.1. Typical cycloid for the wheeled vehicle rolling on peat terrain.

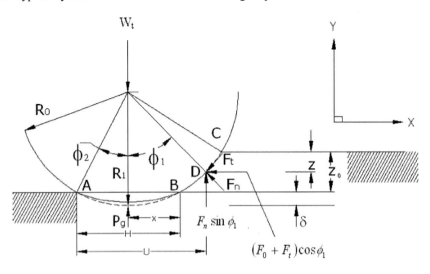

Figure 6.2. Points on the perimeter of a driving wheel describe a looped cycloid.

Figure 6.3 shows a wheel of the vehicle rolls on peat terrain with load and the displacement of the contact point of the wheel relative to the peat terrain. Individual points of the wheel perimeter move along looped cycloid.

$$x = U - H/2 \qquad (6.5)$$

where $U = R_1 \sin \phi_1 + R_1 \sin \phi_2$ and $\phi = \phi_1 + \phi_2$

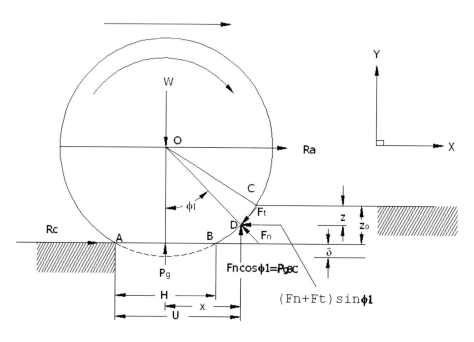

Figure 6.3. Wheel-terrain interaction model.

In Equation (6.5), R_1 is radius of the point on the perimeter, which is being examined, U the length of the contact surface or the horizontal projection of the contact arc, and ϕ the central angle which belongs to U.

It is assumed that the driving wheels operate with exerting tangential force (tractive force) due to the applying driving torque at the wheel, and its vertical deformation and the accompanying soil deformations create a condition as if the wheels roll with slip with radius R_z. The radius R_z can be measured by using the following equation:

$$R_z = R_0 (1 - i_{rd}) \qquad (6.6)$$

where, $i_{rd} = \left(1 - \dfrac{V}{R_o \omega}\right) \times 100\% = \left(1 - \dfrac{R_z}{R_o}\right) \times 100\%$

In Equation (6.6), i_{rd} is the slippage of wheel relative to the terrain in percentage which is induced due to the friction at the interface between driving wheel and terrain, V is the linear speed at the tire wheel in m/s, and ω is the angular speed of the tire in rad/s. The resulting slip radius R_i can be calculated by using the following equation:

$$R_i = R_z.(1 - i) = R_0.(1 - i_{rd})(1 - i) = R_0(1 - i_r)$$ (6.7)

In Equation (6.7), i_r is the resultant slippage,

$$i_r = i + i_{rd} \text{ and } i.i_{rd} \cong 0 \text{ .with } i_r = \left(1 - \frac{V}{R_i\omega}\right) \times 100\ \%$$

When the driving wheel is under slippage, the displacement of the vehicle due to the slippage (i.e, slip displacement) of the wheel can be represented as:

$$x = U - \frac{H}{2} = R_i\phi - \frac{H}{2} = R_0\phi(1 - i_r) - \frac{H}{2}$$ (6.8)

The resultant slippage, i_r is caused by the tangential force and tire deformation. Equation (4.8) represents slip displacement when considering the kinematics interaction at the tire-terrain interface. When the deformable tire rolls with radius R_0, the resultant slippage is zero, or $i_r = 0$. The displacement of the tire can be written as

$$x_0 = R_0\phi - \frac{H}{2}$$ (6.9)

In Equation (6.9), x_0 is the displacement of the tire when $i_r = o$. When $i = 0$, the slip displacement of the tire will be zero ($x = 0$), and Equation (6.9) can be rewritten as

$$x = R_0\phi(1 - i_{rd}) - \frac{H}{2} = 0$$

$$i_{rd} = i_r = 1 - \frac{H}{(2R_0\phi)}$$

$$= \left(1 - \frac{V}{R_z\omega}\right) \times 100\ \%$$

$$= \left(1 - \frac{V}{R_i\omega}\right) \times 100\ \%$$ (6.10)

In this study it is assumed that the vehicle ground contact pressure is higher than the normal ground pressure i.e, $P_{vc} > P_g$, the tire operates in elastic mode, and the lower part of the tire in contact with the terrain is flattened. The inflation pressure of the tire is reduced; it is assumed that the deflection of the tire will be δ. The contact length of the tire H can be computed by considering the vertical equilibrium of the tire. The length of the contact surface is definitely the function of the tire deflection. Therefore, H is computed by considering the tire deflection δ ,

$$H = 2\sqrt{[(\delta)(D - \delta)]}$$
$$= (2R_o\phi)\left(1 - \left(1 - \frac{V}{R_i\omega}\right) \times 100\ \%\right) \tag{6.11}$$

where, D is the diameter of the tire in meter. In Figure 4, the point 'D' is the normal force acting point on the curve BC of the wheel and 'z' is defined as the depth of the normal force acting point (i.e, $z = f(Position\ of\ D)$. The slip sinkage of the vehicle is computed in this study by using $z' = z_o - z$. The slip sinkage is defined as the sinkage of the vehicle due to the slippage. It is earlier mentioned that the slippage of the vehicle increases with stucking the vehicle. If the slip sinkage of the vehicle equals to zero (i.e $z' = 0$), the point 'D' would be at the point B. While, the point 'D' would be at the middle of curve BC if $z = \frac{1}{2}z_o$. Therefore, the entry angle $\phi 1$ and exit angle $\phi 2$ invoked during forward traveling on the terrain, may be computed as:

$$\phi_1 = \sin^{-1}\left(\frac{2x}{D}\right) \tag{6.12}$$

$$\phi_2 = \sin^{-1}\left(\frac{H}{2D}\right) \tag{6.13}$$

$$\phi = \sin^{-1}\left(\frac{2x}{D}\right) + \sin^{-1}\left(\frac{H}{2D}\right) \tag{6.14}$$

It can be further written based upon the geometry shown in Figure 6.3:

$$x = \sqrt{\{D(z_0 + \delta - z)\} - (z_0 + \delta - z)^2}$$

or

$$x = \sqrt{\{(z_0 + \delta - z)(D - (z_0 + \delta - z))\}} \tag{6.15}$$

The assumption for the stress distribution addressed in Ref. [6] that the motion resistance of a rigid wheel is due to the vertical work done in making a rut of depth z_0. The force F_n acts as the normal force on the curve BC as point D of the wheel's tangential force F_t, leading to following equations: If the slip sinkage, $z \cong {z_0}/{2}$

$$x = \sqrt{\left\{ \left({z_0}/{2} + \delta \right) \left(D - {z_0}/{2} - \delta \right) \right\}} \tag{6.16}$$

If slip sinkage, $z \cong z_0$

$$x = \sqrt{[(\delta)(D - \delta)]} \tag{6.17}$$

The total entry and exit angle ϕ of the tire can be computed by simplifying Equations (6.14) and (6.16) as shown in the follows: For slip sinkage, $z \cong {z_0}/{2}$

$$\phi = \sin^{-1} \left(\frac{2\sqrt{\left\{ \left({z_0}/{2} + \delta \right) \left(D - {z_0}/{2} - \delta \right) \right\}}}{D} \right) + \sin^{-1}\left({H}/{2D} \right) \tag{6.18}$$

For slip sinkage, $z \cong z_0$

$$\phi = \sin^{-1} \left(\frac{2\sqrt{\left\{ \delta (D - \delta) \right\}}}{D} \right) + \sin^{-1}\left({H}/{2D} \right) \tag{6.19}$$

The slippage of the vehicle can be computed as follows. For slip sinkage, $z \cong {z_0}/{2}$

$$i_{rd} = 1 - \frac{2\sqrt{\left\{ \delta (D - \delta) \right\}}}{2R_0 \left[\sin^{-1}\left\{ \frac{2\sqrt{\left\{ \left({z_0}/{2} + \delta \right) \left(D - {z_0}/{2} - \delta \right) \right\}}}{D} \right\} + \sin^{-1}\left(\frac{2\sqrt{[\delta (D - \delta)]}}{D} \right) \right]} \tag{6.20}$$

For slip sinkage, $z \cong z_0$

$$i_{rd} = 1 - \frac{2\sqrt{\left\{ \delta (D - \delta) \right\}}}{2R_0 \left[\sin^{-1}\left\{ \frac{2\sqrt{\left\{ \delta (D - \delta) \right\}}}{D} \right\} + \sin^{-1}\left(\frac{2\sqrt{\left\{ \delta (D - \delta) \right\}}}{D} \right) \right]} \tag{6.21}$$

In practice, it would be more convenient to follow an iterative process to determine the values of δ, H and φ by using the Equations (6.11), (6.18), and (6.19) with assuming the values of i_{rd} in the range of 10-100%. Finally, the values of i_{rd} could be verified by the computing values of H and φ.

Load Distribution

The normal load W is supported by the normal ground pressure P_g on the tire flattened portion AB and arc portion BC. Therefore, part of the tire load will be supported by the portion of the curve BC as shown in Figure 6.4. The vehicle load on the curve BC due to the effect of the vertical component of the wheel tangential force $F_t \sin \varphi_1$ is considered to be W_{BC} and it could be computed as follows:

$$W_{BC} = B \int_0^{z_0} P_g \, dx$$

$$W_{BC} = B \int_0^{z_0} \left(k_p z + \frac{4}{D_h} m_m z^2 \right) dx \tag{6.22}$$

The computation of the vehicle load distribution could be made for minimum slip sinkage to critical sinkage. For the minimum slip sinkage of the vehicle W_{BC} could be made by differentiating the Equation (6.22) with neglecting the secondary term and we have,

$$dx = \frac{- D\,dz}{2\sqrt{\{D(z_0 + \delta - z)\}}}$$

Substituting dx in Equation (6. 22) leads to:

$$W_{BC} = \int_0^{z_0} \frac{B\sqrt{D}}{2\sqrt{(z_0 + \delta - z)}} \left(k_p z + 4 m_m \frac{(H+B)}{(HB)} z^2 \right) dz \tag{6.23}$$

Assuming $(z_0 + \delta - z) = p^2$, results in $dz = -2\,p\,dp$, Equation (6.23) can be rewritten as

$$W_{BC} = \int_{\sqrt{(z_0 + \delta)}}^{\sqrt{z}} B \left(k_p z + 4 m_m \frac{(H+B)}{HB} z^2 \right) \left(\sqrt{-D} \right) dp \tag{6.24}$$

By using, $z = \left(z_0 + \delta - p^2 \right)$ in Equation (6.24), we have,

$$W_{BC} = -Bk_p \sqrt{D} \int_{\sqrt{(z_0+\delta)}}^{\sqrt{\delta}} (z_0 + \delta - p^2) dp - \frac{(H+B)}{HB}(4m_m)\sqrt{D} \int_{\sqrt{(z_0+\delta)}}^{\sqrt{\delta}} (z_0 + \delta - p^2)^2 dp \tag{6.25}$$

It is considered that the vehicle load will be supported by the underlying peat stiffness k_p if the vehicle sinkage is more than the surface mat thickness of 0.12 to 0.7m. This study focuses on the effect of vehicle sinkage, $z \leq 0.10\ m$ and the maximum slip sinkage (i.e., $z \cong z_0$). Thus the equation (6.25) can be rewritten as follows:

$$W_{BC} = -Bk_p \sqrt{D} \int_{\sqrt{(z_0+\delta)}}^{\sqrt{z}} (z_0 + \delta - p^2)\, dp \tag{6.26}$$

By simplifying equation (5.26), we can write

$$W_{BC} = Bk_p \sqrt{D} \left[\frac{2}{3}\delta^{1/2}(\delta + 1.5z_0) - \frac{2}{3}(\delta + z_0)^{2/3} \right] \tag{6.27}$$

The vertical force applied on the tire can be expressed as

$$W = BHp_g + W_{BC} \tag{6.28}$$

Traction Force

The traction of the vehicle for the tire-terrain interfaces can be calculated by using the following equation recommended in Ref. [5]:

$$F_T = 2B \int_0^L \tau\, dx \tag{6.29}$$

where F_T is the traction force in kN, B is the width of the track in m, τ is the shearing stress in kN/m^2. The shearing stress of the peat terrain can be determined by using the equation of Wong (1989):

$$\tau = \tau_{max} \left(\frac{j_x}{K_w} \right) \exp\left(1 - \frac{j_x}{K_w} \right) \tag{6.30}$$

where, τ_{max} is the maximum shearing stress in kN/m^2, c is the cohesiveness in kN/m^2, j_x is the slip displacement in m, and K_w is the shearing deformation in m. The maximum shearing stress can be determined as,

$$\tau_{max} = (c + \sigma \tan \varphi)$$

(6.31)

$$j_x = ix$$

(6.32)

From the wheel-terrain interaction model as shown in Figure 6.3, the tractive force of the vehicle could be determined by applying the Newton's motion law:

$$(F_t + F_n)\sin \varphi_1 - R_c - R_a = \frac{W}{g}\frac{d}{dt}(v)$$

$$(F_t + F_n)\sin \varphi = R_c + R_a + \frac{W}{g}\frac{d}{dt}(v)$$

(6.33)

where, $(F_t + F_n)\sin \varphi = 2B \int_0^L \tau dx$ and $F_t + F_n = F_{wp}$

In equation (6.33), F_t is the tangential force in kN, F_n is the normal force that exert from the terrain in kN, W_t is the total of the vehicle weight in kN, R_c is the motion resistance due to terrain compaction in kN and v is the traveling speed of the vehicle in km/h.

By integrating Equation (6.33), the traction equation of the tire-terrain interfaces can be written as,

$$F_{wp} = \frac{1}{(\sin \varphi_1)} A_t (c + \sigma_t \tan \varphi) \left[\frac{K_w}{iL_t} e^1 - \left(1 + \frac{K_w}{iL_t} \right) \exp \left(1 - \frac{iL_t}{K_w} \right) \right]$$

(6.34)

where $\sigma = \dfrac{W_t}{A_t}$ and $A_t = BL_t = B \left(H + x - \dfrac{H}{2} \right) = B \left(\dfrac{H}{2} + x \right)$

In Equation (6.34), F_{wp} is the traction force of the vehicle in the tire-terrain interaction in kN.

Motion Resistance

The motion resistance due to tire deformation can be computed by using the equation recommended in Ref. [4]:

$$R_h = \frac{\left[3.58 BD^2 P_g \varepsilon (0.035 \alpha - \sin 2\alpha) \right]}{\alpha (D - 2\delta)}$$

(6.35)

where $\alpha = \cos^{-1} \left[\dfrac{(D - 2\delta)}{D} \right]$ and $\varepsilon = 1 - \exp \left(- k_e \delta / h \right)$

In Equation (6.35), α is the contact angle in degrees, h the tire section height in m, and k_e a parameter related to tire construction. The values of k_e is 7 for radial tire and 15 for bias-ply tires addressed in ref. [1].

6.3. TRACTIVE PERFORMANCE INVESTIGATION

Vehicle tractive performance in terms of tractive effort and slippages was investigated by simulation and conducting the field experiment. The comparison on the simulation and the field experimental results has been made to substantiate the validity of the developed mathematical model in this study. The wheel vehicle as shown Figure 6.5 could be made suitable to operate on the low bearing capacity peat terrain either by reducing 15% inflated pressure in order to increase the tire-terrain contact length by 40% and contact width by 35% or by reducing the LGP-30 wheeled vehicle total load to 9.81 kN without payload and 19.62 kN with 9.81 kN payload. The field experiments were conducted to verify the traction of the wheel system by reducing 15% inflated pressure in order to increase the tire-terrain contact length by 40% and contact width by 35%.

6.3.1. Tractive Performance Investigation - Theoretically

The suitability of the vehicle as shown in Figure 6.5 is justified with its tractive performance simulation by considering the peat terrain mechanical properties such as moisture content ω, bulk density γ, cohesiveness c, internal friction angle φ, shear deformation modulus K_W, surface mat stiffness m_m, and underlying peat stiffness k_p as stated in Table 6.1.

Figure 6.5. LGP-30 Wheeled vehicle.

The specifications of the LGP-30 vehicle are shown in table 2. The main drive ratio of this vehicle power transmission system is 1:1. For the simulation of the vehicle LGP-30 on

Sepang peat terrain, the vehicle travelling distance is considered to be 200 m, the terrain mean surface mat thickness and underlying peat thickness were considered 0.12 m and 3 m, respectively.

The critical sinkage of the vehicle is considered to be 0.120 m as the surface mat thickness was found 0.12 m from the earlier study reported in Ref. [8]. The field experiment for getting the mechanical properties of Sepang peat has been conducted in 15 different points over the 200 m traveling path of the vehicle and the same experimental methods has been repeated 5 times over the 200 m. The simulation on the LGP-30 vehicle is made on the vehicle sinkage, pressure, slippage, tractive effort (traction), and motion resistance by considering the mechanical properties of the terrain for the 15 different points over the 200 m vehicle traveling path. The sinkage and the tractive equation are quite similar to those recommended in Ref. [9]. The kinematics model in this study is validated with the field experimental results addressed in ref. [10]. It is noted that if the vehicle wheel sinkage is less than 100 mm the vehicle will be supported by both the surface mat and the underlying peat. But, if the vehicle sinkage is 120 mm or more the vehicle is in risk because the vehicle is supported only by the strength of the underlying peat.

Figure 6.6 shows the typical sinkage of the LGP-30 vehicle on the Sepang peat terrain, showing the minimum sinkage of 0.24 m and 0.28m for the 26 kN and 34.5 kN LGP-30 vehicle, respectively. As mentioned previously, most of the vehicle load is supported by the surface mat of the peat terrain, but in the case when the vehicle minimum sinkage is 0.24 m for the vehicle of 24 kN, the load of the vehicle would be supported by the underlying peat only. Furthermore, when the ground clearance of the vehicle is 0.48 m, the vehicle is completely in risk to traverse on the peat terrain unless some modifications are adopted in order to make the vehicle sinkage less than 0.10 m.

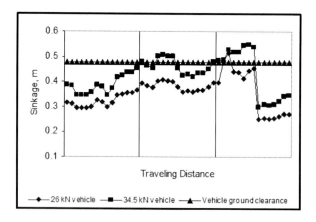

Figure 6.6. Typical sinkage of the LGP-30 wheeled vehicle on Sepang peat terrain.

Figure 6.6 shows that the slippage of the vehicle increases with the increase in vehicle load. Figure 6.7 shows that the vehicle's sinkage and slippage as a function of travel distance. It is noted that the sinkage of the vehicle increases with the increase in vehicle slippage. It could be concluded that the vehicle sinkage is a function of vehicle slippage. Equation 6.20 represents the same relationship as the relationship of the sinkage and slippage in Figure 6.8.

Furthermore, it should be mentioned that if the vehicle slippage increases too much the vehicle will spin rather than rolling, which is the major cause of the vehicle sinkage.

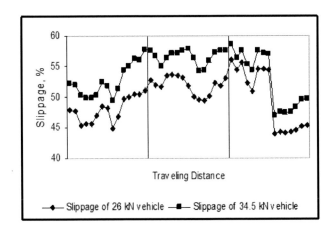

Figure 6.7. Typical slippage of the LGP-30 wheeled vehicle on Sepang peat terrain.

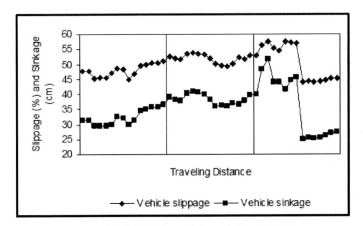

(a) LGP-30, vehicle weight=26 kN

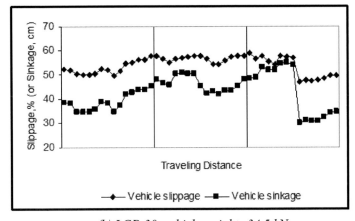

(b) LGP-30, vehicle weight=34.5 kN

Figure 6.8. Relationship between vehicle slippage and sinkage.

This study indicates that the vehicle sinkage for the LGP-30 wheeled vehicle should not be more than 0.48 m as the vehicle ground clearance is 0.48 m and the maximum slippage should not be more than 40% in order to allow the LGP-30 vehicle to move. The mobility of off-road vehicles is justified based on the relationship between vehicle ground contact pressure and ground normal pressure distribution, especially for peat or muskeg terrain. If the ground contact pressure of the vehicle is higher than the vehicle ground normal pressure, the vehicle would not be able to traverse on the terrain rather to sink. Based on Figure 6.9, it could be concluded that the vehicle LGP-30 would not be able to mobile on the terrain as the vehicle normal ground pressure is much higher than the vehicle ground normal pressure.

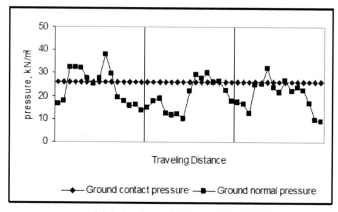

(a) LGP-30, vehicle weight=26 kN

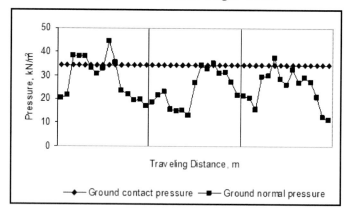

(b) LGP-30, vehicle weight=34.5 kN

Figure 6.9. Pressure distribution of the LGP-30 wheeled vehicle.

Figure 6.10 shows the vehicle tractive force for the slippage of 20%, 30% and 40%. Based on this it could be concluded that the vehicle traction decreases with increasing slippage of the vehicle and the vehicle tractive force increases with increasing the vehicle weight. Figure 9 shows that the maximum tractive force 44.23, 34.62, and 23.85% of the vehicle weight 34.5 kN for the slippage of 20%, 30% and 40%. Figure 10 shows that the motion resistance is significantly higher than the traction of the vehicle, which could happen due to the excessive sinkage of tires.

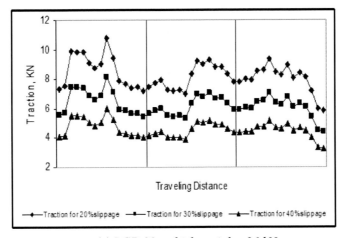

(a) LGP-30, vehicle weight=26 kN

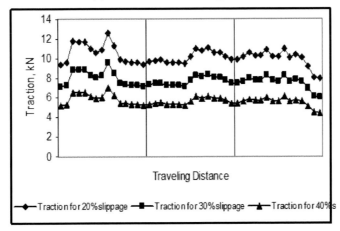

(b) LGP-30, vehicle weight=34.5 kN

Figure 6.10. Traction of the wheeled vehicle on Sepang peat terrain.

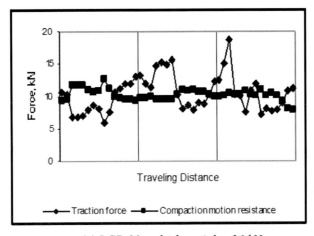

(a) LGP-30, vehicle weight=26 kN

Figure 6.11. (Continued).

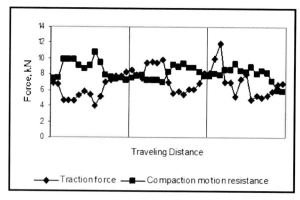

(b) LGP-30, vehicle weight=34.5 kN

Figure 6.11. Relationship between tracion and motion resistance.

Furthermore, the simulation results presented in figure [6.8-6.11], show that the vehicle

- Sinkage is more than the critical sinkage of 0.12 m
- Slippage is more than critical slippage of 40 %
- Ground contact pressure is more than the critical pressure of 17 kN/m^2
- Rolling motion resistance due to terrain compaction is sometimes more than tractive force of the vehicle.

It is thus concluded that the vehicle would not be suitable to traverse on the peat terrain. The vehicle could be made suitable to operate on the low bearing capacity peat terrain either by

- Reducing 15% inflated pressure in order to increase the tire-terrain contact length by 40% and contact width by 35%. The resulted vehicle's ground contact pressure and peat terrain's ground pressure are shown in Figure 6.12.
- Reducing the LGP-30 wheeled vehicle total load to 9.81 kN without payload and 19.62 kN with 9.81 kN payload.

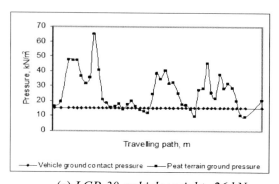

(a) LGP-30, vehicle weight=26 kN

Figure 6.12. (Continued).

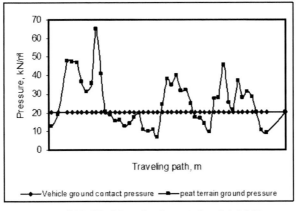

(b)LGP-30, vehicle weight=34.5 kN

Figure 6.12. Pressure relationship after decreasing 15% tyre inflation pressure.

6.3.2. Tractive Performance Investigation: Experimentally

The vehicle testing site was the unprepared moderate peat terrain at the Sepang, opposite of the Kuala Lumpur International Airport (KLIA), Malaysia. The important instrumentation system system are installed on the vehicle for conducting the field experiment on the vehicle tractive performance.

The tractive force of the vehicle was measured by using the torque transducer and the traveling speed was set by using the K3GN-NDC-FLK DC24 digital panel meter. The slippage of the vehicle was conducted by measuring the vehicle actual speed by radar sensor and theoretical speed by electromagnetic-pick-up sensor.

The motion resistance tests were not conducted due to the parking problem of the auxiliary vehicle. The straight motion tests of the vehicle were performed by increasing the vehicle's tyre-terrain contact part (flattened portion) with decreasing 5%, 10% and 15% tire inflation pressure. Before each of the test, the vehicle was made ready by installing the portable generator set and the DEWE 2010 on the vehicle. The instrumentation system was tested by executing the developed programmed with DASY Lab 5.6® into the DEWE-2010.

Then, a preliminary run on the terrains was performed for ensuring the expected function of the instrumentation system of the vehicle. *The field experiments were conducted on all over the field. Test I represented for a complete trips on one track while the Test II for the different track on the same field.*

Typical field experimental results of the 26 kN vehicle are shown in Figure xxx. The traveling speed of the vehicle during field test was considered 12 km/h which is the recommended velocity of the vehicle for plantation (according to the Malaysia Airport Berhad Plantation' Operational Manager).

The following brief discussions have been made based on the field experimental results:

- It is found that the vehicle was stuck frequently when the tyre pressure was maximum as shown in Figure 6.13. But, it was very rarely when the flattened portion

was increased by decreasing the tyre inflation pressure of 5%. While, the vehicle traversed effectively when its inflation pressure was reduced by 15%.
- In the slippage and the tractive force relationship as shown in Figure 6.14, it shows that the vehicle tyre-ground contact flattened incremental effect significantly increases the tractive effort and decreases the slippage.

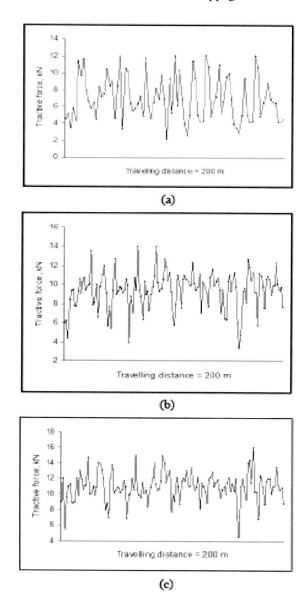

Figure 6.13. 26 kN LGP-30 wheeled vehicle typical tractive force over 200 m traveling distance with keeping constant the tyre-terrain contact part flattened by decreasing the tyre inflation pressure of (a) 5%, (b) 10%, and (c) 15%.

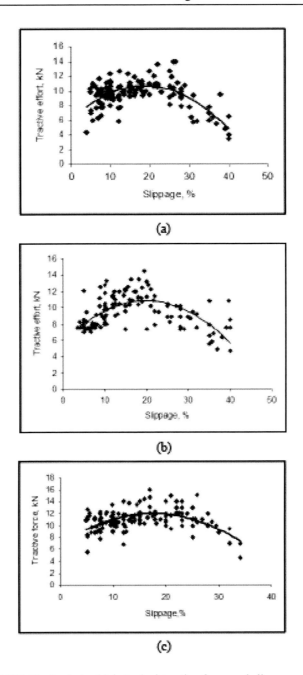

Figure 6.14. 34.5 kN LGP-30 wheeled vehicle typical tractive force and slippage relationship with keeping constant the tyre-terrain contact part flattened by decreasing the tyre inflation pressure of (a) 5%, (b) 10%, and (c) 15%.

This conclusion could be justified by simplifying the Equation (32), which stated that the tractive effort of the vehicle is mainly the function of the vehicle tyre-ground contact area and the slippage. The terrain cohesiveness, internal friction angle, the shear deformation modulus could be not affect the tractive force significantly. While, the vehicle weight is constant. Furthermore, the incremental flattened portion of the tyre increased the vehicle floatation capacity.

6.4. MATHEMATICAL MODEL VALIDITY

The validation of the developed mathematical model in this study was carried out by making comparison of the measured and predicted tractive performance of the vehicle in straight motion for the loading conditions of 26 kN and 34.5 kN with increasing the tyre flattened portion by decreasing the tyre inflation pressure of 15%. To validate the mathematical model, the vehicle tractive performance in terms of tractive force and slippage was measured and compared with the predicted ones.

Figure 6.15 shows the comparison of the predicted and measured tractive force for difference slippage during straight motion for the loading conditions of 26 kN and 34.5 kN. It indicates that the predicted data over the measured data has a closed agreement and thus the closed agreement could substantiate the validity of the mathematical model during straight motion.

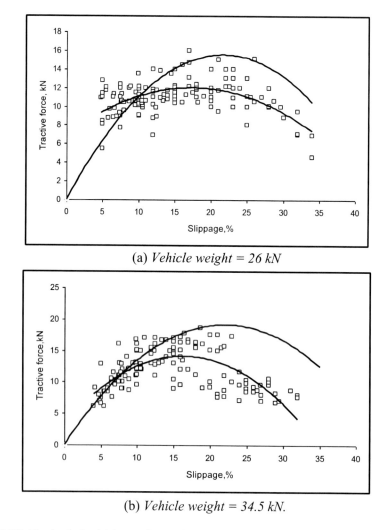

(a) *Vehicle weight = 26 kN*

(b) *Vehicle weight = 34.5 kN.*

Figure 6.15. LGP-30 wheeled vehicle tractive performance comparison keeping the tyre flattened portion constant by decreasing 15% tyre inflation pressure.

The results indicate that there is less variability of the measured data and predicted data of the LGP-30 wheeled vehicle. It was found that the variability of the predicted data over the measured data is 19.2% and 23% for the vehicle loading conditions of 26 kN and 34.5 kN, respectively.

The results indicate that there is less variability of the measured data and predicted data of the LGP-30 wheeled vehicle. It was found that the variability of the predicted data over the measured data is 19.2% and 23% for the vehicle loading conditions of 26 kN and 34.5 kN, respectively. It indicates that the predicted data over the measured data has a closed agreement and thus the closed agreement could substantiate the validity of the mathematical model during straight motion.

CONCLUSION OF THE CHAPTER

Following conclusions could be drawn based on the content of this chapter:

- The new mathematical model on the slippage represents the proper function of the vehicle sinkage.
- Sinkage and slippage increased with increasing of the terrain moisture content. It is noted that excessive sinkage of the vehicle causes the vehicle stuck which increased the vehicle slippage significantly.
- Tractive effort of the vehicle changed over the travelling distance for the variation of the terrain cohesiveness c, internal friction angle φ, and the slippage of the terrain as the vehicle's operating load and the tire contact area are considered to be constant. It is reported in the authors earlier study Ref. [8], the cohesiveness and internal frictional angle are inversely proportional with the terrain moisture content.
- Motion resistance of the vehicle changes with the variation of the terrain surface mat stiffness m_m, underlying peat stiffness k_p and the sinkage as the vehicle operating load and the tire width are constant.
- The simulation result also shows that the maximum tractive effort is 44.23, 34.62, and 23.85% of the vehicle weight for the slippage of 20, 30, and 40%, respectively.
- LGP-30 wheeled vehicle suitability for traversing on the moderate peat terrain could be justified by compromising the vehicle payload and the reduction 15% of the tyre inflation pressure for the tyre-terrain contact flattened part.

REFERENCES

[1] J.Y.Wong.*Theory of Ground Vehicles*, (Third Edition), New York: John Willey and Sons, Inc. 2001.

[2] J.Y.Wong, M. Garbar, and Preston-Thomas.'Theoritical prediction and experimental substantiation of the ground pressure distribution and tractive performance of tracked vehicles'. Journal of Automobile Engineering. *Proceedings of the IMech E Part D*, 15, pp.65-85, 1984.

[3] G. Komandi. 'Reevaluation of the adhesive relationship between the tire and the soil' *Journal of Terramechanics*, 30, pp.77-83, 1993.

[4] M.G. Bekker and E.V. Semonin. 'Motion resistance of pneumatic tires', *Journal of Automotive Engineering*, l.6(2),1975.

[5] M.G. Bekker. Introduction to terrain-vehicle systems. *Ann Arbor*, MI: University of Michigan Press,1969.

[6] R. Ataur, Y. Azmi and A K M Mohiuddin. Mobility investigation of a designed and developed segmented rubber track vehicle for sepang peat terrain in Malaysia'. *Journal of Automobile Engineering.* Proceedings of the IMech E Part D, 221 (D7), pp.789-800, 2007.

[7] R. Ataur, Y. Azmi, B. Zohaide, D. Ahmad, and W. Ishak. Mechanical properties in relation to mobility of Sepang peat terrain in Malaysia. *Journal of Terramechanics*, 41(1), pp.25-40, 2004.

[8] R. Ataur, Y. Azmi, B. Zohadie, D. Ahmad, and W. Ishak. Tractive performance of a designed and developed segmented rubber tracked vehicle on Sepang peat terrain during straight motion: theoretical analysis and experimental substantiation, *Int. J. Heavy Vehicle Systems*, 13(4), pp.298–323,2006.

[9] R. Ataur, Y. Azmi, B. Zohadie, D. Ahmad, and W. Ishak. Design and Development of a Segmented Rubber Tracked Vehicle for Sepang Peat Terrain in Malaysia. *Int. J. of Heavy Vehicle Systems*, Inderscience. UK, 12(3),pp.239-267,2005.

Chapter 7

TRACKED VEHICLE FOR MODERATE PEAT TERRAIN

7.1. INTRODUCTION

Fully mechanized the palm oil on peat terrain is basically difficult by wheeled vehicles. Different types of peat tracked vehicles are introduced and tested but none is achieved to solve the total mechanization problems. Peat Prototype Tracked Tractor with ground contact pressure of 19.75 kN/m^2 as presented by Ooi [2], FEB Picker wheeled vehicle with ground contact pressure of 30.44 kN/m^2 by Mustasim et al. [10], MRK-1 wheeled vehicle with ground contact pressure of 33.8 kN/m^2 by Shuib et al. [3] and MALTRAK tracked vehicle ground contact pressure of 21.5 kN/m^2 by Yahya et al. [4] on the peat terrain for the collection-transportation operations of FFB. It was reported that none of the tracked vehicle or the wheel vehicle configurations that are designed and developed in Malaysia were able to traverse on the low bearing capacity peat. This is true simply because these vehicles were not designed and developed to meet peat terrain requirements. This study introduces a new tracked vehicle with ground contact pressure of 12.69 kN/m^2 which is mainly designed based on the conditions of the low bearing capacity Sepang peat.

The new vehicle proposed in this chapter ensures high mobility over the low bearing peat and is able to collect-transport the FFB under any working conditions. Furthermore, the maintenance cost of the vehicle is decreased by replacing the damaged segment of the track with a new one rather than replacing the entire track.

Moderate peat terrain traficability is very low to prevent the vehicle from sinkage reported by Rahman et al. (2004). The wheeled vehicle should not be considered for the moderate peat terrain as the vehicle train contact area is very low. While tracked vehicle should be considered for the moderate peat teraain as the tracked vehicle contact area can be adjusted to fit for the operation on the terrain.

7.2. TRACTION MECHANICS OF TRACK VEHICLE

The mechanics of wheel systems and rubber tracks in many points of view are similar (Grisso et al., 2006). The vehicle ground pressure distribution is adjustable by selection of suitable ground contact area. The ground contact area is adjustable by varying the track contact length and track contact width (Dwyer et al., 1993). In order to avoid excessive slippage for rubber tracks in situation of soft and organic terrains like muskeg and peat the choice of longer track length is beneficial, although achieving the desirable pressure distribution is subjected to negotiate about the adjustment of ground contact area (Wong, 2008). Zoz (1997) analyzed the general characteristics of traction mechanism of rubber wheeled and rubber tracked tractors and reported that by equipping wider rubber tracks and larger diameter tires we would reach to the greatest improvement on tractive performance under difficult tractive conditions.

For a commonly encountered soft and organic soil similar to low bearing capacity peat terrain in Malaysia, there is a mat of living vegetation on the surface and a layer of saturated peat beneath it. However, when the applied pressure (load) reaches to certain level, the surface mat will fail. Since the saturated peat beneath the surface mat is often weaker than mat, and shows lower resistance, the pressure decreases with an increase of sinkage after the surface mat in broken (Wong, 2008). To employ the bearing capacity of the surface mat and fully utilize its shearing strength for generating the traction, It is necessary to use proper grouser on tracks. The risk of cutting the surface mat will be extremely increased by applying the grouser on the track unless the track slippage is properly limited, as proposed by Bekker (1969).

Track system configuration has significant effect on ground pressure distribution which is coming mainly from configuration of the ground wheels (Gigler, 1993). It has been reported that the ground pressure distribution can be reduced by increasing the number of ground wheels and the ground pressure distribution will reach to a higher uniformity. Increasing the number of ground wheels of track configuration system from 5 to 8 ground wheels would result in an decrease of ground pressure distribution from 9.25 to 5.08 (Wong and Preston-Thomas, 1988). Wong et al. (2008) compared three vehicles of same weight and dimensions but different track configuration systems to find the effect of track configuration on vehicle mobility. Vehicle A has 5 ground wheels and the torsion bar suspension system and the other two were the same as vehicle A but one with six ground wheels and next with eight overlapping ground wheels, vehicle A(6W) and Vehicle A(8W) respectively. Ogorkiewicz (1991) reported that the overlapping configuration like vehicle A (8W) was widely used by Germany during World War II. The overlapping configuration will results in more uniform pressure distribution compare to the other configurations and also results in less track sinkage and reduces track motion resistance and consequently leads to better trafficability but it must be mentioned that it may also cause increasing of design complexity and manufacturing cost and extra weight of the vehicle which is not desirable for low bearing capacity peat terrain (Wong, 2008). Furthermore it must be noted that on peat terrain, which there is cohesive soil, the spacing between overlapping ground wheels maybe clogged with mud and terrain materials therefore causing operational problems. Wong (2008) reported that if the track initial tension is too low, the track will be loose, hence the track segments are unable to support much load between two consecutive ground wheels and track segments right under the ground wheels are responsible to carry the substantial weight of the vehicle.

The mathematical model for the vehicle engine power and tractive performance computations was made based on vehicle's straight and turning motion with non-uniform pressure distribution. The non-uniform ground pressure distribution of the vehicle was achieved by locating the centre of gravity (CG) at the rearward of the mid-point of the track ground contact length. The vehicle ground pressure distribution was assumed to increase from front idler to the rear sprocket.

The mathematical model was developed by simplifying the general tractive equations and motion resistance equations of Wong et al. [12], Wong [13], Muro [8], and Okello et al. [10] for peat terrain. In developing the mathematical model for the vehicle for straight motion and turning motion, the track was assumed to be a medium pitch rigid link track. The tractive effort of the vehicle for non-uniform ground pressure distribution during straight motion is developed based on Figure 7.1. It shows that the tractive effort of the vehicle is not only developed on the ground contact part of the track but also on the side parts of the ground contact track grouser.

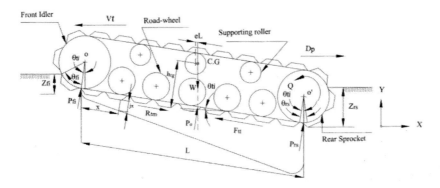

Figure 7.1. Force acting on tracked system.

Furthermore, it is not only developed on the main part of the ground contact track but also on parts of the front idler and rear sprocket. The general equations for computing the tractive effort of the vehicle during straight and turning motion are stated in the following sections.

Consider a rigid link segmented rubber track vehicle of total weight W, track size including track ground contact length L, width B, pitch T_p, and grouser height H, radius of the front idler R_{fi}, rear sprocket R_{rs}, and road-wheel R_w, and height of the center of gravity h_{cg} is traversing under traction on a peat terrain at a constant speed of v_t as soon as applying driving torque Q at the rear sprocket.

The pressure distribution in the track-terrain interface is assumed to be uniform by locating the vehicle C.G at the track mid point of the track system. The track entry and exit angle at the front idler θ_{fi} and rear sprocket θ_{rs} reveal same values for the same diameter of the front idler and rear sprocket. The reaction pressures at the front idler P_{fi}, main straight part P_o, and rear sprocket P_{rs}, and the sinkage of the front idler z_{fi}, main straight part z_{mp}, and rear sprocket z_{rs} reveal also same value due to the location of the C.G at the middle of the track system as shown in Figure 7.1.

7.2.1. Straight Motion

The tractive effort of the vehicle is computed by using the equation of Rahman et al. 2006.

Under the bottom of the track:

$$F_b = A \left(c + \sigma \tan \varphi\right) \left[\frac{K_w}{iL} e - \left(1 + \frac{K_w}{iL}\right) \exp\left(1 - \frac{iL}{K_w}\right) \right]$$

(7.1)

with $A = 4\left(B \times L\right)$, $\sigma = \dfrac{W}{A}$

$$\sigma = \sigma_{fi} + \sigma_{ms} + \sigma_{rs}, \quad L = L_{fi} + L_{ms} + L_{rs}, \text{ and } \quad i = i_{fi} + i_{ms} + i_{rs}$$

where, F_b is the tractive effort that develop at the bottom part of the track in kN, A is the contact area of the track in m^2, c is the terrain cohesiveness in kN/m^2, σ is the vehicle normal stress in kN/m^2, σ_{fi} is the normal stress acts on the bottom part of the front idler in kN/m^2, σ_{ms} is the stress on the main straight parts in kN/m^2, σ_{rs} is the stress on the bottom parts of the rear sprocket in kN/m^2, φ is the terrain internal friction angle in degree, K_w is the shear deformation modulus in m, L is the ground contact part of the track in m, L_{fi} is the length of contact part of the front idler in m, L_{ms} is the contact part of the main straight part in m, L_{rs} is the contact part of the rear sprocket in m, i is the slippage of the track in percentage, i_{fi} is the slippage of the front idler, i_{ms} is the slippage of the main straight part, i_{ms} is the slippage of the rear sprocket.

The slippage of the ground contact track of front idler can be represented by the following derived equations:

$$i_{fi} = \left(\frac{R_{fi}}{L_{fi}}\right) \int_0^{\theta_{fi}} \{1 - (1 - i)\cos(\theta + \theta_{ti})\} d\theta$$

(7.2)

By integrating the equation (6.2), the slippage of the front idler can be computed as,

$$i_{fi} = \left(\frac{R_{fi}}{L_{fi}}\right)\left[\frac{\pi}{180}(\theta_{fi}) - (1 - i)\{\sin(\theta_{fi} + \theta_{ti}) - \sin\theta_{ti}\}\right]$$

where $L_{fi} = \left(R_{fi}\right)\left(\theta_{fi} + \theta_{ti}\right)$

Similarly, the slippage of the rear sprocket can be computed as,

$$i_{rs} = \left(\frac{R_{rs}}{L_{rs}}\right)\left[\frac{\pi}{180}(\theta_{rs}) - (1 - i)\{\sin(\theta_{rs} + \theta_{ti}) - \sin\theta_{ti}\}\right]$$

(7.3)

where $\quad L_{rs} = R_{rs}\left(\theta_{rs} + \theta_{ti}\right), \quad \theta_{ti} = \arcsin\left(\dfrac{R_{rs} - z_{rs}}{L}\right), \quad \theta_{fi} = \arccos\left(\cos\theta_{ti} - \dfrac{z_{fi}}{R_{fi}}\right)$ and

$$\theta_{rs} = \arcsin\left(\frac{z_{rs} - z_{fi}}{L}\right) + \arccos\left[\frac{R_{rs}}{\sqrt{\left\{R_{rs}^{2} + \left(z_{rs} - z_{fi}\right)^{2}\right\}}}\right]$$

where, θ_{ti} is the track belt trim angle with ground in degree, θ_{fi} is the front idler entry angle in degree, θ_{rs} is the rear sprocket exit angle in degree, z_{fi} is the front idler sinkage in m and z_{rs} is the sinkage of the rear sprocket in m, R_{fi} is the front idler radius in m and R_{rs} is the rear sprocket radius in m.

The slippage of the main track straight part can be computed by using the following equation:

$$i_{mp} = \frac{i_{fi} + i_{rs}}{2} \tag{7.4}$$

Under the **side** of the track:

$$F_{s} = 4HL\left(c + \sigma\tan\varphi\right)\cos\alpha\left[\frac{K_{w}}{iL} e - \left(1 + \frac{K}{iL}\right)\exp\left(1 - \frac{iL}{K_{w}}\right)\right] \tag{7.5}$$

with $\quad \alpha = arc\ \cot\left(\dfrac{H}{B}\right)$

where, F_{s} is the thrust developed to the side of the front idler grouser in kN, H is the height of the grouser in m, and α is the angle of the track system between grouser and width in degree.

The motion resistance of the vehicle due to terrain compaction can be represented by the following derived equation of Rahman et al(2007):

$$R_{c} = \left(2B\right)\left[\begin{array}{l}\left[\left(\dfrac{L_{fi}}{L}\right)\left(\dfrac{k_{p}z_{fi}^{2}}{2} + \dfrac{4}{3D_{hfi}} m_{m}z_{fi}^{3}\right) + \left(\dfrac{L_{mp}}{L}\right)\left(\dfrac{k_{p}z_{mp}^{2}}{2} + \dfrac{4}{3D_{hmp}} m_{m}z_{mp}^{3}\right)\right] \\ + \left(\dfrac{L_{rs}}{L}\right)\left(\dfrac{k_{p}z_{rs}^{2}}{2} + \dfrac{4}{3D_{hrs}} m_{m}z_{rs}^{3}\right)\end{array}\right] \tag{7.6}$$

where, $D_{hfi} = \dfrac{2BL_{fi}}{\left(L_{fi} + B\right)}$, $D_{hmp} = \dfrac{2BL_{mp}}{\left(L_{mp} + B\right)}$, and $D_{hrs} = \dfrac{2BL_{rs}}{\left(L_{rs} + B\right)}$

R_{c} is the motion resistance of the evhicle due to terrain in compact in kN, B is the track width in m, z_{fi}, z_{mp}, and z_{rs} are the sinkages of the vehicle in m, D_{hfi}, D_{hmp}, and D_{hrs} are the hydraulic diameters of the front idler track, track main straight part, and track of rear of sprocket, respectively. k_p is the internal peat stiffness in kN/m^3, and m_m is the surface mat stiffness in kN/m^3.

The sinkage of the front idler, main straight part and rear sprocket can be represented by the equation. of Rahman et al., (2006):

$$z_{fi} = \frac{-\left(\dfrac{k_p D_{hfi}}{4m_m}\right) \pm \sqrt{\left[\left(\dfrac{k_p D_{hfi}}{4m_m}\right)^2 + \dfrac{D_{hfi} P_{fi}}{m_m}\right]}}{2} \quad , \quad z_{ms} = \frac{z_{fi} + z_{rs}}{2}$$

and

$$z_{rs} = \frac{-\left(\dfrac{k_p D_{hrs}}{4m_m}\right) \pm \sqrt{\left[\left(\dfrac{k_p D_{hrs}}{4m_m}\right)^2 + \dfrac{D_{hrs} P_{rs}}{m_m}\right]}}{2}$$

where, P_{fi} is the pressure under the bottom part of the front idler in kN/m^2 and P_{rs} is the pressure under the bottom part of the rear sprocket in kN/m^2.

The ground pressure distribution of the vehicle tracked-terrain interfaces during loading and unloading can be represented by using the equation of Muro (2005):

$$P_{fi} = P_0\left(1 - 6e_i\right) \tag{7.7}$$

where $e_i = \dfrac{20}{L}$ when $e_i \angle \dfrac{1}{6}$

$$P_{rs} = P_0\left(1 + 6e_i\right) - P_{fi} - P_u \tag{7.8}$$

where, P_0 is the normal pressure that exit from the vehicle in kN/m^2, P_u is the unloading pressure in kN/m^2, and e_i is the load eccentricity.

7.2.2. Traction Mechanics for Turning Motion

Figure 2 shows that the effective driving tractive effort F'_{ot} acting on the outer track and the effective braking or driving tractive effort F'_{it} acting on the inner track can be represented from the balance of acting forces on each of the tracks during turning on terrain with radius R and a speed of 10 km/hr as follows:

For driving both the tracks,

$$F_{tt} = F'_{ot} + F'_{it} = \left(F_{ot} - R_{\ln ot}\right)\cos\beta + \left(F_{it} - R_{\ln it}\right)\cos\beta \tag{7.9}$$

where $F_{ot} = F_{lot} + F_{sot}$ and $F_{it} = F_{lit} + F_{sit}$

For driving outer track and braking inner track,

$$F_{tt} = \cos \beta \left(F_{ot} - R_{\ln ot} \right) \tag{7.10}$$

where, Ftt is the effective tractive effort of the vehicle in kN, Fot and Fit are the outer and inner track tractive effort in kN, Flot and Flit are the longitudinal tractive effort and Fsot and Fsit are the tractive effort of the side of the track, Rlnit and Rlnot are the longitudinal motion resistance for the inner and outer track, and β is the slip angle for the vehicle in degree. The tractive effort of the vehicle for longitudinal movement can be represented by the following equation of Rahman et al. [2006]:

$$F_{Lo\,(i)t} = LB \left(c + \sigma_{o(i)t} \tan \varphi \right) \left[\frac{K_w}{i_{o(i)t} L} e^1 - \left(1 + \frac{K_w}{i_{o(i)t} L} \right) \exp \left(1 - \frac{i_{o(i)t} L}{K_w} \right) \right] \tag{7.11}$$

where $\sigma_{o(i)t} = \dfrac{W_{o(i)t}}{LB}$

where, $F_{Lo(i)t}$ is tractive effort that develops at the bottom part of the outer or inner track in kN, L is the ground contact part of the track in m, $\sigma_{o(i)t}$ is the vehicle normal stress either for either outer track or inner track in kN/m^2, c is the terrain cohesiveness in kN/m^2, φ is the terrain internal friction angle in degree, K_w is the shear deformation modulus in m, and $i_{o(i)t}$ is the slippage of the vehicle in percentage.

As a consequence of the shifting of the center of turn, the equivalent moment of turning resistance M_r shows two components: one is the moment of lateral resistance exerted on the tracks by the terrain about O′ and the other is the moment of the centrifugal force about O′. Thus, the moment of turning resistance M_r about O′ can be computed by using the equation of Rahman et al. (2006):

$$M_r = \frac{W_{ot} \mu_L}{L} \int_0^{L/2 + D - C_x} x dx - \frac{W_{it} \mu_L}{L} \int_0^{L/2 - D + C_x} x dx + \frac{W \Omega^2 R}{B_{stc} g} D \cos \beta - \frac{W \Omega^2 R}{B_{stc} g} C_x \sin \beta$$

$$= \frac{\mu_L}{2L} \left[W_{ot} \left(\frac{L}{2} + D - C_x \right)^2 - W_{it} \left(\frac{L}{2} - D + C_x \right)^2 \right] + \frac{W \Omega^2 R}{B_{stc} g} (D \cos \beta - C_x \sin \beta) \tag{7.12}$$

with

$$R = \frac{B_{stc}}{2} \left[\frac{g(1 - i_{ot}) + (1 - i_{it})}{g(1 - i_{ot}) - (1 - i_{it})} \right] \cos \beta \quad , \quad \Omega = \frac{R_{rs} \left[\omega_{ot}(1 - i_{ot}) - \omega_{it}(1 - i_{it}) \right]}{B_{stc}} \cos \beta$$

$$W_{ot} = \frac{W}{2} + \frac{h_{cg}W\Omega^2 R}{B_{stc}g}\cos\beta \; , \; W_{it} = \frac{W}{2} - \frac{h_{cg}W\Omega^2 R}{B_{stc}g}\cos\beta$$

and $\vartheta = Speed\;ratio = \dfrac{\omega_{ot}}{\omega_{it}}$

where, $F_{Lo(i)t}$ is tractive effort that develops at the bottom part of the outer or inner track in kN, L is the ground contact part of the track in m, $\sigma_{o(i)t}$ is the vehicle normal stress either for either outer track or inner track in kN/m^2, c is the terrain cohesiveness in kN/m^2, φ is the terrain internal friction angle in degree, K_w is the shear deformation modulus in m, and $i_{o(i)t}$ is the slippage of the vehicle in percentage.

As a consequence of the shifting of the center of turn, the equivalent moment of turning resistance M_r shows two components: one is the moment of lateral resistance exerted on the tracks by the terrain about O' and the other is the moment of the centrifugal force about O'. Thus, the moment of turning resistance M_r about O' can be computed by using the equation of Rahman et al. (2006):

$$R_{co(i)t} = (B)\left[\begin{array}{l}\left(\dfrac{L_{fi}}{L}\right)\left(\dfrac{k_p z_{fio(i)t}^2}{2} + \dfrac{4}{3D_{hfi}}m_m z_{fio(i)}^3\right) + \left(\dfrac{L_{mp}}{L}\right)\left(\dfrac{k_p z_{mpo(i)t}^2}{2} + \dfrac{4}{3D_{hmp}}m_m z_{mpo(i)t}^3\right) \\ + \left(\dfrac{L_{rs}}{L}\right)\left(\dfrac{k_p z_{rso(i)t}^2}{2} + \dfrac{4}{3D_{hrs}}m_m z_{rso(i)t}^3\right)\end{array}\right] \qquad (7.13)$$

where, $R_{co(i)t}$ is the total motion resistance of the vehicle due to soil compaction for either outer track or inner.

The lateral motion resistance force exerted on the track by the displacement of the terrain surface can be computed by the derived equation,

$$R_{lot} = \frac{W_{ot}\mu_L}{B_{stc}L}\int_0^{L/2+D-C_x} x\,dx + \frac{W_{ot}\mu_L}{L}\int_0^{L/2-D+C_x} x\,dx$$

$$= \frac{W_{ot}\mu_l}{2B_{stc}L}\left[\left(\frac{L}{2}+D-C_x\right)^2 + \left(\frac{L}{2}-D+C_x\right)^2\right] \qquad (7.14)$$

$$R_{lit} = \frac{W_{it}\mu_L}{B_{stc}L}\int_0^{L/2+D-C_x} x\,dx - \frac{W_{it}\mu_L}{L}\int_0^{L/2-D+C_x} x\,dx$$

$$= \frac{W_{it}\mu_l}{2B_{stc}L}\left[\left(\frac{L}{2}+D-C_x\right)^2 - \left(\frac{L}{2}-D+C_x\right)^2\right] \qquad (7.15)$$

where, R_{lot} and R_{lit} are the lateral resistance of the outer and inner track in kN, W is the total weight of the vehicle in kN, B_{stc} is the track center to center distance in m, L is the track ground contact length in m, μ_L is the lateral motion resistance co-efficient, x is the small segmented length of the track in m, and C_x is the longitudinal distance between C.G and lateral centerline of the vehicle in hull in m.

It is noted that the vehicle lateral resistance must be higher or equal to the vehicle centrifugal force $\left(i.e., R_{lot} + R_{lit} \geq \left(F_{cent} = \left(Wv_t^{\ 2} \cos \beta / gR\right)\right)\right)$ in order to maintain the vehicle stability during turning. Where, v_t is the theoretical velocity in m/s and g is the acceleration due to gravity in m/s^2.

The effective sprocket power can be represented by the following equation:

$$P_{rs} = \frac{QN_{rs}}{9550} \qquad (7.16)$$

where, P_{rs} is the sprocket power in kW, Q is the torque of the sprocket in N-m, and N_{rs} is the speed of the sprocket in rev/min.

The effective engine power available at the pump's input shaft for developing the desired output torque at driven sprockets can be represented by the following equation of Wong (2001):

$$P_e = \left(\frac{1}{367.2}\right)\frac{F_{tt}v_t}{\mu_t} \qquad (7.17)$$

where, P_e is the engine power in kW, F_{tt} is the total tractive effort in kg, μ_t is the efficiency of the hydrostatic transmission system in percentage.

Traction coefficient of the vehicle is defined as the ratio of the vehicle total tractive effort and the vehicle total weight. It is represented by the following equation:

$$\Gamma = \frac{F_{tt}}{W} \qquad (7.18)$$

where, Γ is the traction coefficient in percentage, F_{tt} is the total tractive effort in kN, and W is the weight of the vehicle in kN.

The motion resistance coefficient of the vehicle can be defined as the ratio of the vehicle total motion resistance and the vehicle total weight. It is represented by the following equation:

$$\rho = \frac{R_{tm}}{W} \qquad (7.19)$$

where, ρ is the motion resistance coefficient in percentage, R_{tm} is the total motion resistance coefficient in kN, and W is the vehicle weight in kN.

The drawbar power is referred to as the potential productivity of the vehicle, that is, the rate at which productive work may be done. It is computed using the following equation:

$$P_d = \left(\frac{1}{367.2}\right)(D_p v_a) \qquad (7.20)$$

with $D_p = F_{tt} - R_{tm}$

where, P_d is the drawbar power of the vehicle in kW, D_p is the drawbar pull in kN, and v_a is the actual speed of the vehicle in km/h.

Tractive efficiency is used to characterize the efficiency of the vehicle in transforming the engine power to the power available at the drawbar. It is defined by the following equation:

$$\eta_t = \frac{P_d}{P_e} \qquad (7.21)$$

where, η_t is the tractive efficiency of the vehicle in percentage, P_d is the drawbar power in kW, and P_e is the effective power of the engine in kW.

4. TRACTIVE PERFORMANCE INVESTIGATION

The track prime mover was equipped with a dedicated instrumentation system to measure, display and record in real-time the travelling speed and sinkage of the vehicle during operation. This complete system is made up of a National Instrument cRIO-9004 CompactRIO Real-time Controller Unit (RCU), a National Instrument TPC 2106T Touch Panel Control (TPC), a Trimble AG132 GPS antenna and receiver unit, a Dlink DIR-655 router, a Panasonic TOUGHBOOK CF19 wireless and extremely-rugged laptop, a Honda Eu20i 2.0kVA portable generator set, three units of Omron E4PA-LS200 M1 N ultra-sonic sensors, and an AC/DC power distribution box.

Figure 7.3 shows the simplified block diagram of the completed instrumentation system which has been installed on the vehicle. Proposed vehicle's tracked system has been tested with a 26 kN tracked vehicle as shown in Figure 7.4, running at 10 km/h over a low bearing capacity 12 kN/m^2.

Before each of the test, the vehicle instrumentation system was demonstrated for checking the individual instrument activation.

The instrumentation system was tested by executing the developed programmed with lab View into the CompactRIO on the field, and the speed was set to the K3GN-NDC-FLK DC24 digital panel meter for getting the expected travelling velocity of the vehicle.

Then, a preliminary run on the terrains was performed for ensuring the expected function of the instrumentation system of the vehicle.

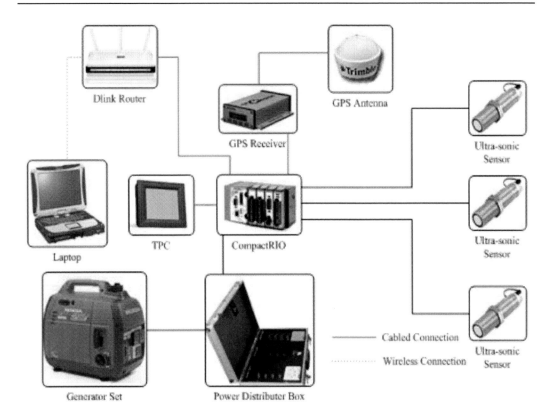

Figure 7.3. Simplified Block Diagram of Complete Instrumentation.

Table 7.1. Technical Specifications of the Tracked Prime Mover

Engine :	
- Model	YANMAR 4TNE84-SAMF
- Type	Vertical, Liquid cooled diesel
- Number of Cylinders	4 Cylinder
- SAE Net continuous power at 3000 rpm, kW and hp (SAE J1349, ISO3046/1Net)	37.7 kW (50.6 hp)
Track :	
- Model	Tonly Endless Metal Cord and Metal Reinforced Rubber Track
- Pitch of Track	90 mm
- Width of Track	350 mm
- Number of pieces	63
Main pumps :	
- Brand	SAMHYDRAULIK
- Model	Variable Displacement
- Series	HCV 50
- Type	Axial Piston
- Displacement	50 cm^3/rev
- Maximum speed	4000 rpm
- Nominal pressure	350 bar

Table 7.1. (Continued)

Engine :	
Motors :	
- Brand	Sai
- Series	GM2-600
- Peak Power	59 kW
- Continuous Speed	450 rpm
- Maximum Speed	700 rpm
- Continuos pressure	250 bar
- Peak Pressure	300 bar
- Torque	8.83 Nm/bar
- Displacement	565 cm^3/rev
Auxiliary pump :	
- Brand	Salami
- Model	Fixed displacement
- Series	Salami 11.3
- Type	Gear pump
- Displacement	11.6 cm^3/rev
- Maximum Speed	4000 rpm
- Nominal Pressure	350 bar
- Prime mover overall dimensions:	
- Length	2800 mm
- Width	2250 mm
- Height	1600 mm

Figure 7.4. Tracked vehicle.

7.4.1. Tractive Performance Investigation - Experimentally

The straight motion tests of the vehicle were performed at two different travelling speeds of 6 km/h and 10 km/h and at two different loading conditions of 12.0 kN and 20.0 kN. Meanwhile, the turning motion tests were performed at a single speed of 16 km/h and at two different loading conditions of 12.0 kN and 20.0 kN. For each of the loading conditions and travelling speeds the vehicle was running two times in a series of travelling paths on each of the terrains. It is to be noted that the engine can produce sufficient torque to the sprocket driven motor only in the range of 2000-2500 rpm. Due to the overheating problem of the engine, the engine was run only at 2000 rpm for the vehicle to complete the turning manoeuvrability at a speed of 16 km/h. Before each of the tests, the vehicle was made ready by installing the portable generator set Honda Eu20i 2.0kVA on the vehicle. The instrumentation system was tested by executing the developed programme with cRIO-9004 CompactRIO Real-time Controller Unit (RCU) on the field, and the speed was set to the K3GN-NDC-FLK DC24 digital panel meter. Then, a preliminary run on the terrains was performed to ensure the expected function of the instrumentation system of the vehicle.

Figure 7.5 shows the typical variation of tractive effort with time during straight motion of the vehicle on the *terrain*. Table 7.2 shows that the mean value of tractive effort of the vehicle significantly differs from one terrain to the other terrain: Terrain III differed from Terrain II by 0.84 kN, Terrain III differed from Terrain I by 1.53 kN, and Terrain II differed from Terrain I by 0.69 kN. The significant difference of the tractive effort between Terrain II and Terrain I is mainly due to the difference of cohesiveness. However, the significant difference between Terrains III and I is mainly due to the differences of cohesiveness and the hydrodynamic effect.

Figure 7.6 shows the typical variation of vehicle tractive effort during turning motion on terrain III. The left track of the vehicle was considered as the outer track and the right track was considered as the inner track. The negative tractive effort of the inner track indicates that the inner track was rotated opposite to the direction of the outer track for few seconds. From the field experimental (measured) data on torque it is found that the outer track sprocket developed the maximum torque of 2133 Nm at the beginning of the turn. Both of the tracks were run afterwards. However, the outer track was faster than the inner track.

Figure 7.7 and 7.8 showed the variation of slippage of the vehicle when on the *Terrain Type* III at traveling speed of 6km/h and 10km/h without payload and with full payload. Figure 7.7 showed that the maximum slippage of the vehicle decreased from 15.2% to 13.68% (decrement of 11.11%) when the vehicle traveling speed increased from 6 to 10km/h. Similarly, Figure 7.8 showed that the maximum slippage of the vehicle decreased from 15.04% to 14.45% (decrement of 4.08%) when the vehicle traveling speed increased from 6 to 10km/h.

Based on the Figures 7.6-7.9, it could be concluded that the tractive effort of the vehicle increased due to the decrease in slippage. The slippage of the vehicle on the *Terrain Type* III is higher at lower speed than at higher speed because of the tearing off the surface of the terrain. This is due to the fact that the tearing off the terrain surface increased the shear displacement of the vehicle on the terrain which could decrease the vehicle tractive effort and increase the vehicle slippage of the vehicle (Radforth, 1958). But, the traction coefficient of the vehicle decreased from 55% to 44% (decrement of 25%) when the vehicle loading condition changed from without payload to with full payloadfor the traveling speed of 6km/h.

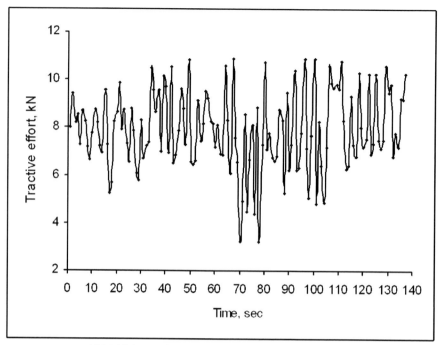

(a) Vehicle tractive effort at travelling speed of 6km/h

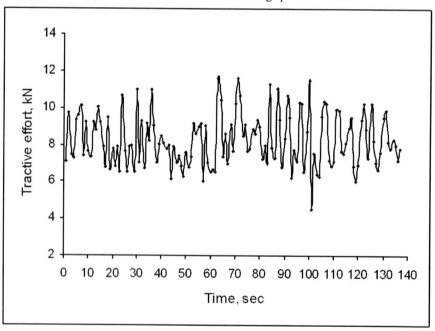

(b) Vehicle tractive effort at traveling speed of 10km/h

Figure 7.6. Vehicle tractive effort on Terrain Type III during straight running motion without payload.

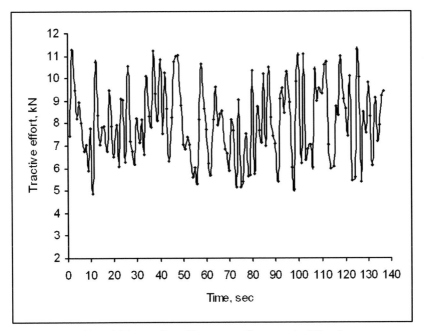

(a) Vehicle tractive effort at traveling speed of 6km/h

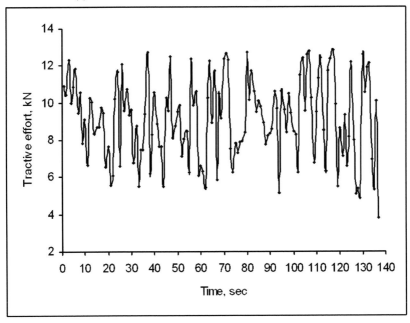

(b) Vehicle tractive effort at traveling speed of 10km/h

Figure 7.7. Vehicle tractive effort on Terrain Type III during straight running motion with full payload.

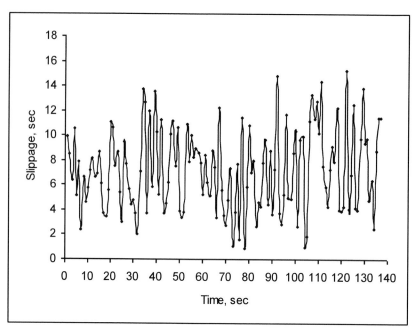

(a) Vehicle slippage at traveling speed of 6km/h

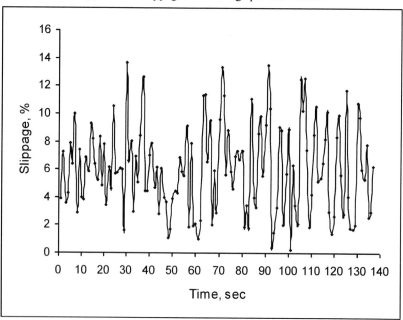

(b) Vehicle slippage at traveling speed of 10km/h

Figure 7.8. Vehicle slippage on the Terrain Type III during straight running motion without payload.

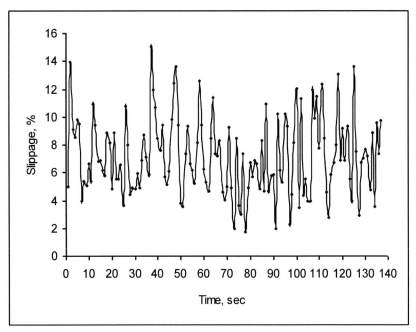

(a) Vehicle slippage at traveling speed of 6km/h

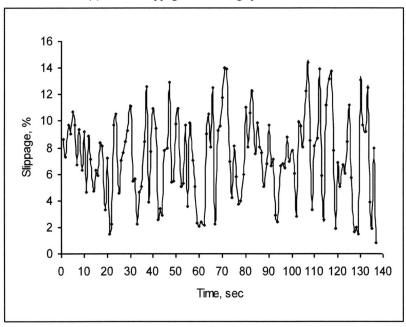

(b) Vehicle slippage at traveling speed of 10km/h

Figure 7.9. Vehicle slippage on the Terrain Type III during straight running motion with full payload.

Furthermore, the traction coefficient decreased from 59.37% to 50.31% (decrement of 18%) when the vehicle's loading condition changed from without payload to with full payload for the traveling speed of 10km/h. The reason of this situation happened could be different from the *Terrain Type* I and *Terrain Type* II. The reason for the increase of tractive effort could be due to the higher rate of loading. For the peat with a higher moisture content, the higher the rate of loading the greater the peat reaction will be. This is due to the hydrodynamic effect arising from the water movement within the peat (MacFarlane, 1969, and Wong et al., 1988).

7.4.2. Result Interpretation: Statistical Analysis

The statistical analysis of this study was designed as a 3×2 factorial where the factors were *terrain*s (Terrain I, Terrain II, Terrain III), and vehicle travelling speeds of 6 km/h and 10 km/h. The statistical design analysis was repeated for the 12.0 kN and 20.0 kN vehicles. Duncan LSD test was conducted to differentiate the mean value of the vehicle tractive effort and slippage for each of the levels of terrain, loading conditions (12.0 kN and 20.0 kN) and velocities. The tested parameters in the statistical analysis were considered to be highly significant only if the probability level is less than 0.01 (i.e., Pr < 0.01) and significant only if the probability level is less than 0.05 (i.e., Pr < 0.05).

Table 7.3 shows that the mean effects of velocity and terrain have a high significant effect on vehicle tractive effort at a probability level less than 0.01 (i.e., Pr<0.01). Meanwhile, the mean effects of the interaction of velocity*terrain have a significant effect on the tractive effort of the vehicle at a probability level less than 0.05 (i.e., Pr<0.05).

The significant effect of velocity indicates that the tractive effort of the vehicle is increased with increasing vehicle travelling velocity. Such a situation occurs due to the increase of the terrain strength and slip-sinkage at higher rates of loading. Furthermore, the strength of the terrain increases and slip-sinkage decreases due to the hydrodynamic effect arising from the water movement within the peat. MacFarlane [15] and Wong et al. [12] mentioned that the values of strength parameters of peat or muskeg obtained at low rate of loading are lower than those obtained at a high rate of loading.

The lower rate of loading decreased the peat strength parameters which are governed by the rate of pore water pressure dissipation. It is also found from the direct field experiment on Sepang peat with the traversing vehicle.

Table 7.3. Analysis of variance of vehicle tractive effort

Source	DF	Sum of Squares	F Value	Pr > F
Velocity	1	67.29	21.69	0.01**
Terrain	2	59.37	9.57	0.01**
Velocity*Terrain	2	27.03	4.36	0.013*
Error	816	2531.31		
Corrected Total	821	2685.01		

** Highly significant at probability level 1 % and *Significant at probability level 5%

The significant effect of the terrain indicates that the tractive effort of the vehicle is increased with the increasing of moisture content of the selected terrain. It is noted that the tractive effort of the vehicle is directly related with the terrain cohesiveness, vehicle weight and the terrain internal frictional angle Bekker [14]. Wong et al. [12] mentioned that a noticeable portion of the vehicle tractive effort is derived from the cohesion of the terrain. It is due to the fact that the terrain cohesiveness and internal frictional angle could cause the variation of the vehicle tractive effort as the vehicle weight is fixed. The significant effect of velocity*terrain interaction indicates that the velocity and terrain react differently to the tractive effort of the vehicle.

Table 7.4 shows the variations of the tractive effort of the vehicle with changes in the vehicle's travelling velocity. For the vehicle of 12.0 kN: the results show that the vehicle tractive effort increases by 13.71% for Terrain I, 11.09% for Terrain II and 13.53% for Terrain III with increasing the traveling velocity of the vehicle from 6 to 10 km/h. This situation occurred mainly by increasing the strength of the terrain and increasing the rate of loading. Wong et al. [12] mentioned that the higher the rate of loading the greater the peat reaction will be.

This is due to the hydrodynamic effect arising from the water movement within the peat. The results also show that the mean value of the tractive effort of the vehicle at a traveling speed of 6 km/h increases by 8.08%, 5.12%, and 14.14% when the vehicle changed its operating environment from Terrain I to Terrain II, Terrain II to Terrain III and Terrain I to Terrain III, respectively. Similarly, for the vehicle travelling speed of 10 km/h, the tractive effort of the vehicle increases by 6.32%, 7.42%, and 14.22% when the vehicle changed its operating environment from Terrain I to Terrain II, Terrain II to Terrain III and Terrain I to Terrain III, respectively. This happened because of the variation of cohesiveness as the vehicle weight remained constant. For the vehicle of 20.0 kN: the results show that the vehicle's tractive effort increases by 16.66% for Terrain I, 6.42% for Terrain II and 15.21% for Terrain III by increasing the traveling velocity of the vehicle from 6 to 10 km/h.

The results also show that the mean value of the tractive effort of the vehicle increased by 15.32%, 6.42%, and 22.74% when the vehicle changed its operating environment at a travelling speed of 6 km/h from Terrain I to Terrain II, Terrain II to Terrain III and Terrain I to Terrain III, respectively. Similarly, the tractive effort of the vehicle increased by 5.22%, 15.2%, and 21.22% when the vehicle changed its operating environment at a travelling speed of 10 km/h from Terrain I to Terrain II and Terrain II to Terrain III, and Terrain I to Terrain III, respectively. The reason for this situation is also similar as discussed in Table 3.

Table 7.5 shows the analysis of variance of the vehicle's tractive effort during turning motion. For the outer track: the result shows that the mean effects of loading conditions and terrain has a significant ($Pr<0.01$) effect on the tractive effort of the vehicle. For the inner track: the result shows that the mean effects of loading conditions have a significant ($Pr<0.05$) effect on the tractive effort. The significant effect of load indicates that the vehicle is able to develop more tractive effort by adding more loads. Wong [13] mentioned that the forward tractive effort of the vehicle is limited by the terrain properties.

Therefore, the added load must be in the range of the terrain bearing capacity. The significant level of terrain indicates that the forwarding tractive effort of the vehicle during turning varies from terrain to terrain as the terrain properties and the hydrodynamic effect arising on the respective terrain are different due to the presence of moisture content.

Table 7.6 shows the variation of the vehicle's track tractive effort during turning manoeuvres. For the outer track, the result shows that the tractive effort of the vehicle outer track increases by 18.89% for Terrain III, 13.29% for Terrain II, and 22.97% for Terrain I when the vehicle loading condition changes from 12.0 kN to 20.0 kN. The variation of tractive effort on each of the terrain was due to increasing the load as the terrain properties for each of the respective terrain type are same. The maximum percentage of tractive effort variation was found for Terrain I because of the lowest moisture content. Furthermore, the slippage of the vehicle of 12.0 kN is higher than the slippage of the vehicle of 20.0 kN on Terrain I and this is due to the lower value of cohesiveness. For the inner track, the result shows that the tractive effort of the vehicle increases by 9.20% for Terrain III, 4.58% for Terrain II, and 11.68% for the Terrain I when the vehicle loading condition changes from a vehicle without a payload to a vehicle with a full payload. The variation of the tractive effort on each of the terrains was due to the increasing load.

Table 7.4. Variations of the vehicle tractive effort on different terrains

Velocity (km/h)	Tractive effort (kN)					
	Terrain I		Terrain II		Terrain III	
	Mean	SD	Mean	SD	Mean	SD
	Vehicle weight=12.0 kN					
6	6.49	1.26	7.43	1.76	7.43	1.76
10	7.38	1.91	8.43	1,81	8.44	1.81
Vehicle weight = 20.0 kN						
6	6.44	1.64	7.43	1.76	7.90	1.81
10	7.51	1.81	7.90	1.81	9.11	2.19

Table 7.5. Analysis of variance on of the 20.0 kN vehicle's inner track tractive effort in turning motion

Source	DF	Sum of Squares	F Value	Pr > F
Outer track				
Loading condition	1	62.71	52.25	0.0001**
Terrain	2	33.37	13.90	0.0001*
Loading *Terrain	2	3.931	1.64	0.1959
Error	354	424.86		
Total	359			
Inner track				
Loading condition	1	7.00	2.93	0.05*
Terrain	2	1.95	0.41	0.67
Load*Terrain	2	2.55	0.53	0.59
Error	354	846.07		
Total	359			

**Highly significant at probability level 1%, *significant at probability level 5%

Table 7.6. Variations of the 20.0 kN vehicle tractive effort in turning motion

Load	Tractive effort (kN)					
	Terrain I		Terrain II		Terrain III	
	Mean	SD	Mean	SD	Mean	SD
	Outer track					
12.0 kN Vehicle	4.98	0.95	5.87	0.92	5.87	0.91
20.0 kN Vehicle	6.12	1.01	6.65	1.07	6.51	1.61
Inner track						
12.0 kN Vehicle	2.31	1.3	2.4	1.56	2.39	1.56
20.0 kN Vehicle	2.58	1.47	2.51	1.63	2.61	1.71

7.5. Chapter Conclusion and Recommendation

7.5.1. Conclusions

The development of the vehicle and the mechanical properties of the Sepang peat terrain were presented. Developing mathematical models descriptions of the track forces and motions were presented. A new design parameters optimization technique to optimize the design parameter of the vehicle was presented. Important instrumentation system was installed on the vehicle to measure the tractive performance. The measured tractive performances were compared with the reported tractive performance to evaluate the vehicle design potentiality. The measured tractive performances were also compared with the predicted tractive performance to valid the developing mathematical models.

1. Nominal ground pressures of a 25.5kN rubber tracked vehicle including total payload of 5.89kN on 2x320x2120 mm^2 track are 20.24kN/m^2 under un-drained conditions and 25.8kN/m^2 under drained conditions whereas the vehicle ground contact pressure is 18.79kN/m^2. The vehicle ground contact pressures are 7.78% and 37.30% lower than the nominal ground pressure of the vehicle for un-drained and drained conditions, respectively.

2. Nominal ground pressure of a 25.5kN pneumatic wheel vehicle on 11.00R16XL low ground pressure tyres at 200kPa inflation pressures are 35.31kN/m^2 under un-drained conditions and 48.069kN/m^2 under drained conditions whereas the normal ground pressure of the vehicle is 63.77kN/m^2. The vehicle ground contact pressures are 195% and 140.74% higher than the nominal ground pressure of the vehicle under un-drained and drained conditions, respectively.

3. Bearing capacity for Sepang peat terrain under un-drained was found to be 18.89kN/m^2 and lower than the vehicle nominal ground pressure for the two considered vehicle configurations. Thus any vehicle could only traverse on Sepang peat terrain without bogging down problem if the vehicle ground contact pressure is less than the bearing capacity of the terrain.

4. Nominal ground pressures of the track vehicle and the pneumatic wheel vehicle under un-drain condition were found to be 18.66kN/m^2 and 68.84kN/m^2, respectively. Thus based on the earlier argument, only the track vehicle could traverse under the un-drained Sepang peat terrain.

5. The mathematical models were developed with understanding the track-terrain interaction mechanism and vehicle traveling speed to measure the vehicle engine power requirement and to simulate the vehicle tractive performance both in straight and turning motion. The following conclusions were drawn based on the developed mathematical models:
   - The developed mathematical models for straight motion can be used to simulate the vehicle tractive performance over the peat terrain for any type of vehicles.
   - The developed mathematical models for turning motion can be used to simulate the vehicle tractive performance over the peat terrain for only tracks vehicle.

- The proper engine size for the vehicle on peat terrain can be determined by using these developed mathematical models and the identified optimize design parameters.
- Simulated maximum slippage, maximum motion resistance to total vehicle weight ratio, minimum drawbar to total vehicle weight ratio and minimum tractive efficiency for the vehicle with full payload during straight motion at the allowable maximum sinkage of 100mm were 15%, 8.9%, 32.2% and 72%, respectively.
- Simulated tractive efficiency for the vehicle with full payload of having 7 road wheels and ratio of road-wheel spacing to track pitch 2.25 is 4.76% higher than the Wong (2001) indicated tractive efficiency for track vehicle having 7 road wheels and ratio of road-wheel spacing to track pitch 2.25 traversing on clayey soil.

6. Based on the vehicle tractive performance analysis during turning, the vehicle is to be designed to have a turning radius of at least 4m under a turning speed of 16km/h for the better turning ability and turning stability. While, the vehicle without payload can turn at the turning radius within the 1.0m to 7.0m range to maintain a steady state turn.

7. The effects on tractive performance of individual design parameters, such as the size of the track, ratio of the road-wheel spacing to track pitch, the ratio of the sprocket pitch diameter to track pitch were described. This provided the basic for the optimization of the overall design of the track system for the Sepang peat terrain. The following conclusion were made based on the optimization design parameters of the vehicle:
- The optimize track ground contact area of $320x2120mm^2$ allow the vehicle sinkage 10% less than the peat terrain surface mat thickness and exit ground pressure of 1.0% less than Sepang peat terrain bearing capacity for un-drained condition.
- The optimize track ground contact area of $320x2120mm^2$ provide the vehicle motion resistance coefficient of 6.54%.
- The optimize grouser height ensures the vehicle to fully utilize the shear strength of the surface mat and to increase the trafficability of the terrain as the surface mat of the Sepang was found to be 40% higher than the grouser height.
- The optimize sprocket pitch diameter of 400mm and location at the rear part of the track system configuration provide the vehicle less sinkage, optimum speed fluctuation and optimum tractive efficiency.
- The optimum front idler diameter of 400mm provides the vehicle of higher track entry angle, lower slippage and optimum tractive efficiency.
- The optimum ratio of the road-wheel spacing to track pitch of 2.25 provides the vehicle to develop traction coefficient of 36%.

8. The vehicle was successfully designed and developed for traversing on peat terrain as a prime mover for infield collection-transportation operation. The main special feature of the vehicle track system are: replacement segmented rubber track which could simplify servicing activities and reduce maintenance cost; longer wheelbase

and shorter road-wheels interval with track system gave better vehicle floatation, reduce vehicle sinkage, and increase vehicle traction.

9. Tractive performance of the vehicle was evaluated by testing the vehicle on three different peat terrains *Terrain Type* I, *Terrain Type* II, and *Terrain Type* III with two different loading conditions and two different speeds. Tractive performance of the vehicle was evaluated mainly based on tractive effort and slippage.

10. Tractive effort of the vehicle at traveling speed of 6km/h increased 2.28% for *Terrain Type* I, 5.124% for *Terrain Type* II, and 6.46% for *Terrain Type* III when the vehicle changing operating loading condition from without payload to with full payload. Similarly, the tractive effort of the vehicle at traveling speed of 10km/h increased 1.76% for *Terrain Type* I, 2.61% for *Terrain Type* II, and 6.69% for *Terrain Type* III when the vehicle changing operating loading condition from without payload to with full payload.

11. Slippage of the vehicle at traveling speed of 6km/h increased 8.15% for *terrain type* I, 11.95% for *terrain type* II, and 17.17% for *terrain type* III when the vehicle changed its operating loading condition from without payload to with full payload. Similarly, vehicle slippage at traveling speed of 10km/h decreased 7.18% for *terrain type* I, increased 3.98% for *terrain type* II, and increased 20.15% for *terrain type* III when the vehicle changed its operating loading condition from without payload to with full payload.

12. Vehicle design could be considered optimum as the measured tractive effort of the vehicle was found to be 36% of the vehicle gross weight which is in the recommended tractive effort range of 30 to 36% of the vehicle gross weight as suggested by Wong (2001).

13. Less variability of the vehicle tractive effort for straight motion in the range of 7.5 to 13.2% between the predicted and measured tractive effort on the peat *terrain type* III for the different loading conditions and operating speeds substantiate the validity of the model.

14. Less variability of the tractive effort for turning motion in the range of 9% to 11.5% between the predicted and measured tractive effort on the peat *terrain type* III for the different loading conditions substantiate the validity of the model.

7.5.2. Recommendations

The following recommendations are made in order to further stabilize the developed mathematical models and the vehicle development:

1. It is important to test the motion resistance of the vehicle on the terrain type III for evaluating the motion resistance coefficient, drawbar pull and tractive performance of the vehicle on the peat terrain.

2. Further field testing is needed to provide a valid comparison between the measured and predicted tractive performance of the vehicle.

3. This study was limited on the mobility of the vehicle not on the productivity. The vehicle productivity can be evaluated from further study with mounting a fruit picking unit and a 1000 kg fruit container.

4. Front part of the vehicle is recommended to be made as a boat shape by fixing a strong stainless steel plate which could be able to prevent the vehicle from the stuck by roots of the existing trees.

5. Vehicle is powered by a NISSAN TD27 single turbo water cooled high speed and high torque diesel engine. There was not a problem for the vehicle with the engine to run for straight motion. But, it was a rigorous problem for turning motion. It is noted that the engine had to run with high speed in the range of 2000rpm to 2500rpm for turning operation which overheated the engine very fast. Therefore, the vehicle can be made suitable for the operation on the peat terrain in all working conditions by replacing present engine with a high torque low speed engine.

7.6. VEHICLE LIMITATIONS

The simulation of the vehicle tractive performance both in straight and turning motion were made based on the mechanical properties of the peat terrain and developed mathematical models. The vehicle development was made with taking accounts the design parameters that were identified from the design parameters optimization technique. The following limitations were made based on the developed mathematical models, terrain traficability, vehicle suitability.

1. Mathematical models that were developed for computing the sinkage, tractive effort, and external motion resistance of the vehicle during straight motion can be used only for evaluating or computing the sinkage, tractive effort, and motion resistance of vehicle on peat terrain.

2. Mathematical models that were developed for computing the tractive performance of the vehicle during turning motion can be used only for all types of track vehicle.

3. Simulation model that was developed for optimizing design and operating parameters of the vehicle can also be used for different off-road vehicle design and operating parameters optimization.

4. Simulation study was limited to 20% of the vehicle slippage for getting the tractive performance of the vehicle in terms of vehicle sinkage, external motion resistance, drawbar pull and tractive efficiency.

5. Vehicle normal load is limited by the strength of the terrain. Experimentally, it is found that the dynamic load of the vehicle limited by the terrain is 25.86kN if the vehicle is equipped with the track ground contact area of $2120 \times 320 mm^2$. Therefore, the normal load of the vehicle should not be more than 25.86kN.

6. Vehicle operating speed during straight running must be 10km/h in order to meet the off-road vehicle mobility and to increase vehicle productivity.

7. Vehicle ground contact pressure must not exceed $18.89 kN/m^2$ in order to traverse on the low bearing capacity peat terrain without bogging down problem.

8. Avoiding from overheating the engine and the hydraulic oil the operating speed of the vehicle engine must not be exceeded 2500rpm.

9. Vehicle track tension must be fixed at 2.35kN for avoiding the track deflection between two consecutive roadwheels, reducing sinkage, reducing the vibration, and increasing the vehicle tractive performance.

10. Critical sinkage of the vehicle could not be more than 100mm as the surface mat thickness of Sepang peat terrain 100mm, which was considered the supporting pan of the vehicle.

11. Vehicle turning radius must be 4.0m for the higher floatation ability, better turning ability and turning stability.

12. Turning speed of the vehicle must not be less than 10km/hr in order to avoid the load transferring from outer track to inner and to avoid the shearing-off the surface mat.

13. Torque of the individual hydraulic motor output shaft should be in the range of 1000Nm to 2200Nm for developing the sufficient power to the sprocket to propel the vehicle track system both in straight and turning motion and to prevent the vehicle from overturning which might cause due to turning moment resistance.

NOTATION

A	contact area of the track	i	slippage of the vehicle
B	width of the track	K_w	terrain shear deformation modulus
B_{tk}	vehicle tracked tread	L	track ground contact length
c	terrain cohesiveness	M_t	turning moment of the vehicle
C_x	longitudinal distance between c.g and lateral centerline of the vehicle in hull	Q	torque of the sprocket
D	instantaneous centre point shifting distance	R_{init}	longitudinal motion resistance for the inner track
e^l	exponential	R_{inot}	longitudinal motion resistance for the outer track
F_b	tractive effort at the bottom of the track	W	weight of the vehicle
F_s	tractive effort at the side of the track	W_{it}	weight transfer to the inner track
F_{it}	tractive effort of the inner track	W_{ot}	weight transfer to the outer track
F_{ot}	tractive effort of the outer track	x	absissa
F_{Lit}	longitudinal tractive effort of the inner track	α	angle of the track system between grouser and width
F_{Lot}	longitudinal tractive effort of the outer track	β	slip angle
F_{sit}	tractive effort at the side of the inner track	ϕ	terrain internal frictional angle
F_{sot}	tractive effort at the side of the outer track	ρ	normal stress
g	acceleration due to gravity	μ_λ	co-efficient of lateral resistance
H	height of the grouser		

REFERENCES

[1] Dwyer, M. J, Okello, J.A, and Sacrlett, A. J. 1993. A Theoretical Investigation of Rubber Tracks for Agriculture. *Journal of Terramechanics*, 30(4), pp.285-298.

[2] Godbole, R., Alcock, R., and Hettiaratchi, D. 1990. The prediction of tractive performance on soil surfaces. *Journal of Terramechanics*, 31(6), pp.443-459.

[3] Muro, T. 1989 Tractive performance of a bulldozer running on weak ground. *Journal of Terramechanics*, 26(3/4), pp.249-273.

[4] Muro, T, Brien, J.O, Kawahara, S and Tran, D.T. 2001. An optimum design method for robotic tracked vehicles operating over fresh concrete during straight line motion. *Journal of Terramechanics*. Vol.38, pp.99-120.

[5] Okello, J.A., Watany, M., and Crolla, D.A. 1998. Theoretical and Experimental Investigation of Rubber Track Performance Models, *Journal of Agricultural Engineering Res.*Vol. 69, pp.15-24.

[6] Rahman, Ataur., Azmi, Y., Zohadie, M., Ahmad, D and Ishak. 2006, Tractive performance of a designed and developed segmented rubber tracked vehicle on Sepang peat terrain during straight motion: theoretical analysis and experimental substantiation. *Int. J. Heavy Vehicle Systems*, Inderscience Publisher Vol. 13, No. 4, pp.298-323.

[7] Rahman, Ataur., Azmi, Y., Zohadie, M., Ahmad, D. and Ishak, W. 2006(a). 'Traction mechanics of the designed and developed segmented rubber track vehicle for Sepang peat terrain in Malaysia during turning motion: theoretical and experimental analysis', *Int. J. Heavy Vehicle Systems,* Vol. 13, No. 4, pp.324–350.

[8] Rahman, Ataur., Azmi, Y., Zohadie, M., Ahmad, D and Ishak. 2006(b). Performance Investigation of the Designed and Developed Segemented Rubber Tracked Vehicle for Sepang Peat Terrain in Malaysia. Vol.13, No.1, pp. *Int. J of Heavy Vehicle System*, Inderscience Publisher.

[9] Wong, J. Y. J., Radforth, R., and Preston-Thomas, J. 1982. Some further studies on the mechanical properties of muskeg in relation to vehicle mobility. *Journal of Terramechanics*, 19(2), pp.107-127.

[10] Wong, J.Y. 1998. Optimization design parameters of rigid-link track systems using an advanced computer aided method. *Proc. Inst. Mech. Engrs*, Part D, Vol.212, pp.153-167.

[11] Shiller, Z., Serate, W., and Hua, M. 1993.Trajectory planning of tracked vehicles. In *IEEE Transaction on Robotics and Control*, Vol 3, pp.796–801.

[12] Wong, J.Y. 2001. *Theory of Ground Vehicles*, (Third Edition), New York: John Willey & Sons, Inc.

[13] Yong, R.N, Fattah, E.A, and Skiadas, N. 1984. Vehicle traction mechanics. Developments in Agricultural Engineering, Elsevier science, Amsterdam, Netherlands, Vol.3, pp.234-248.

[14] ASAE. 1996. *Agricultural Engineers Yearbook of Standards*, American Society of Agricultural Engineers, Michigan.

[15] Bekker, M.G. 1969. *Introduction to terrain-vehicle systems*. Ann Arbor, MI: University of Michigan Press.

[16] Bekker, M.G. 1975. Tracks in muskeg, *Specialnotiser SFM,* Sweden, No.17.

[17] Bodin, A. 1999. Development of a tracked vehicle to study the influence of vehicle parameters on tractive performance in soft terrain. *Journal of Terramechanics*, 36, pp.167-181.

[18] Dwyer, M. J, Okello, J.A, and Sacrlett, A. J. 1993. A Theoretical Investigation of Rubber Tracks for Agriculture. *Journal of Terramechanics,* 30(4), pp.285-298.

[19] Erbach, D.C, Melvin, S.W, Cruse, R.M and Jannsen, D.C. 1989. Effects of tractor tracks during secondary tillage on corn production. *ASAE paper* 89-1533.

[20] Godbole, R., Alcock, R., and Hettiaratchi, D.1990. The prediction of tractive performance on soil surfaces. *Journal of Terramechanics*, 31(6), pp.443-459.

[21] Jamaluddin, B. J. 2002. Sarawak peat agricultural uses. http://www. Alterra.dlo.nl/. Date ofaccess. Feb 2005.

[22] Kitano, M and Kuma, M. 1977. An analysis of horizontal plane motion of tracked vehicles. *Journal of Terramechanics*, 14(4), pp.21-25.

[23] Okello, J.A., Watany, M., and Crolla, D.A. 1998. Theoretical and Experimental Investigation of Rubber Track Performance Models, *Journal of Agricultural Engineering Res.*Vol. 69, pp.15-24.

[24] Ooi, H.S. 1993. Theoritical investigation on the tractive performance of MALTRAK on soft padi soil. MARDI, Report no.116.

[25] Ooi, H.S. 1996. Design and development of peat prototype track type tractor. MARDI, Report no.184.

[26] J.Y.Wong. *Theory of Ground Vehicles*, (Third Edition), New York: John Willey & Sons, Inc. 2001.

[27] J.Y.Wong, M. Garbar, and Preston-Thomas.'Theoritical prediction and experimental substantiation of the ground pressure distribution and tractive performance of tracked vehicles'. Journal of Automobile Engineering. *Proceedings of the IMech E Part D*, 15, pp.65-85, 1984.

[28] G. Komandi. 'Reevaluation of the adhesive relationship between the tire and the soil' *Journal of Terramechanics*, 30, pp.77-83, 1993.

[29] Rahman,. Ataur, Y. Azmi and A K M Mohiuddin. Mobility investigation of a designed and developed segmented rubber track vehicle for sepang peat terrain in Malaysia'. *Journal of Automobile Engineering. Proceedings of the IMech E Part D,* 221 (D7), pp.789-800, 2007.

[30] R. Ataur, Y. Azmi, B. Zohaide, D. Ahmad, and W. Ishak . Mechanical properties in relation to mobility of Sepang peat terrain in Malaysia. *Journal of Terramechanics*, 41(1), pp.25-40, 2004.

[31] R. Ataur, Y. Azmi, B. Zohadie, D. Ahmad, and W. Ishak. Tractive performance of a designed and developed segmented rubber tracked vehicle on Sepang peat terrain during straight motion: theoretical analysis and experimental substantiation, *Int. J. Heavy Vehicle Systems*, 13(4), pp.298–323,2006.

[32] Rahman, Ataur, Y. Azmi, B. Zohadie, D. Ahmad, and W. Ishak. Design and Development of a Segmented Rubber Tracked Vehicle for Sepang Peat Terrain in Malaysia. *Int. J. of Heavy Vehicle Systems*, Inderscience. UK, 12(3), pp.239-267, 2005.

Chapter 8

INTELLIGENT TRACKED VEHICLE FOR WORST PEAT SWAMP

8.1. INTRODUCTION

The "Hover craft" is and air - cushion vehicle running on two engines of 18.65 kW air cooled two cylinders petrol engine for lifting, and 130.55 kW liquid cooled four cylinder petrol engine for propulsion which is developed by MPOB. The MPOB air cushion vehicle is severely prone to damage by the roots of oil palm trees which scatters all over the field as there is no protection of the air cushion system. Furthermore while moving on the field the machine scatters the loose fruits all around the ground and these machines did not show any success in Malaysian oil palm plantation. An intelligent air cushion tracked vehicle with novel design cushion protection system has been developed. The vehicle performance was investigated. It was found that the vehicle was not in any risk on damaging the cushion system.

Intelligent air-cushion tracked vehicle (IACTV) has become one of the most viable alternatives to conventional vehicles such as fully tracked vehicles and fully wheeled vehicles on the swamp peat terrain. The intelligent air-cushion system (IACS) increases the vehicle floatation capacity and allows the vehicle to traverse over the terrain without any interruption. Though a wide range of human endeavors in such fields as agriculture, oil industry, construction, mining, and military operations still involves locomotion using specialized off-road vehicles, the greater emphasis on tractive performance of tracked vehicle has stimulated a renewed interest in many parts of the world at present and therefore, needs to increase the knowledge about intelligent air-cushion system to ensure the vehicle cross country mobility with increasing the vehicle floatation capacity. The performance of air-cushion tracked vehicles travelling in a straight motion with uniform ground pressure distribution is well understood. However, prepared or unprepared tracks inherently have uneven profile for situations of vehicles travelling on deformable road surfaces. The vehicle responses during off-road operation are dependent on the road conditions and vehicle parameters such as cushion clearance height, vehicle speed, vehicle weight, and air-cushion pressure (Bekker, 1969; Wong, 2008).

Transportation efficiency is the main requirement for the tracked vehicles, in the situation of vehicles travelling over low bearing capacity terrain. Tracked vehicles provide ground pressure of 12 kN/m^2 and have the capability of operating over a wide range of unprepared terrain (Rahman et al., 2004). However, the swamp peat terrain is the most critical terrain in Malaysia where the plantation companies are expanding their plantations reported by Malaysia Agriculture Research Development Institute (MARDI). Jamaluddin (2002) has reported that the swamp peat terrain surface mat thickness is closed to 70 mm and bearing capacity is 7 kN/m^2. Systematic studies of the principles underlying the transportation development of off-road vehicles on low bearing capacity swamp peat terrain, therefore, have attracted considerable interest to the field of agriculture, construction, and the military. Different research institutes such as universities, government organizations, and private companies have proposed and developed different types of tracked vehicles for transportation operation but only for moderate terrain (Ooi, 1996; Bodin, 1999; Ataur et al., 2005). Consequently, it is considered to be very difficult to develop any working machine that could run and operate on swamp peat terrain. In addition, the agricultural developments on the swamp peat are mainly restricted by its low trafficability and high water table.

A semi-tracked air-cushion vehicle (STACV) which combines air-cushion technology with a driving mechanism has been developed and the prototype experiments have been performed in the soil-bin made in the Laboratory (Luo et al., 2003). The soil used in testing is sandy loam. However, the use of commercial air-cushion tracked vehicles to test the vehicle parameters is limited due to the difficulties in varying parameters as well as the control of the air-cushion pressure. In order to get reasonably good performance on low bearing capacity swamp peat terrain, intelligent air-cushion tracked vehicle (IACTV) is chosen for this application.

Transportation operation on swamp peat terrain are greatly emphasized to design and develop the vehicles with high crossing ability, good tractive performance and maneuver on swamp peat terrain. Many research works have been carried out and different types of prototypes on tracked vehicle on low bearing capacity terrain have been introduced by Yong et al. (1984), Okello et al. (1998), Muro et al. (1998), Bodin (1999), Wong (2001), Luo et al. (2003), and Ataur et al. (2005). However, most of the vehicles are wheeled vehicles, tracked vehicles and air-cushion semi-tracked vehicles. It is found that the wheel vehicle system is not effective in soft terrain rather it is efficient on hard dry and flat or slightly sloping areas. Among the locally made track vehicles, the Super Crawler by Holy Drilling Machines Sdn. Bhd. has been developed for working efficiently during the wet season when the ground is in soft and worst conditions. It is found that the vehicle is unable to operate on peat terrain due to excessive surface shearing-off which increases the motion resistance. Yahya et al. (1997) has developed the FFB Picker MRK-1 to mechanize and automate the infield collection-transportation of fresh fruit bunches (FFB). But the main problem is that the tracks of the vehicle easily slip out from the traction wheels in turning motion. Segmented rubber track vehicle has been developed by Ataur et al. (2005) for collection-transportation operation on Sepang peat terrain. The ground pressure distributions of these vehicles are found as more than 16 kN/m^2 and none of these vehicles are suitable in terms of their ground pressure distributions with respect to the low bearing capacity of the swamp peat terrain. Luo et al. (2003) has developed the semi-tracked air-cushion vehicle for soft terrain where the vehicle is fixed in such a way that it always slides on the ground with the movement of the vehicle which may result excessive power consumption and increases the vehicle maintenance costs.

The semi-tracked air cushion vehicle of Luo et al. (2003) is not protected from the external threat on the ground by any protecting system and thus, it could be restricted due to the presence of submerged and undecomposed materials such as woods, stones, stumps, and shrubs. The air-cushion system of this study is protected by using a novel-designed supporting system. It adjusts the air-cushion system on the terrain automatically by absorbing its longitudinal displacement with two horizontally attached springs and vertical displacement with four vertically attached springs. Furthermore, the air-cushion system on the developed intelligent air-cushion tracked vehicle (IACTV) is fixed just few millimeters top of the ground that it is used when it is needed. The air-cushion-terrain interaction takes place in an uncertain and vague environment. It could be difficult to describe such a complex mechanics with developing a precise mathematical model. The theoretical modelers who want to derive behavior rules of general nature about the air-cushion-terrain interaction are bound to make simplifying assumptions (George and Maria, 1995). Furthermore, the use of mathematical models with control parameters in air-cushion system for tracked vehicle operating on swamp terrain which involve various types of uncertainties and vague phenomena raises the question how accurately they reflect reality. Hence it is natural to look for alternative methodologies. In this regard, the air-cushion system for an intelligent air-cushion tracked vehicle has been more precisely controlled by applying fuzzy logic controller (FLC) during operation.

7.2. TRACTION MECHANICS

The developed vehicle will be a custom-built which main function will be the transportation operation of agricultural and industrial goods on swamp peat. The development of the vehicle will be made with a comprehensive understanding of the swamp peat terrain physical strength while the intelligent air-cushion system will be developed with taken into account the vehicle dynamics over the terrain.

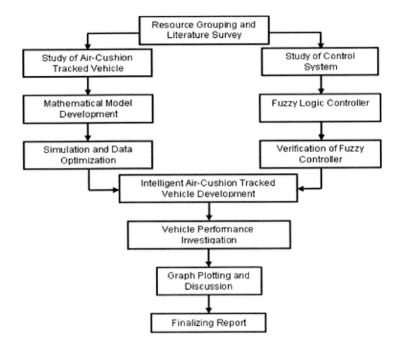

Figure 8.1. Entire methodology of the study.

The mathematical model which will be formulated by understanding the terrain nature, analyzing the mechanics of track-terrain interaction and the interaction of air-cushion-terrain with simplifying mathematical models found in literature. The basic purpose of the mathematical model will to optimize vehicle design parameters and to analyze theoretically its performance without and with air-cushion system.

The mathematical model will be then used for simulating the intelligent air-cushion system performance, optimizing the design parameters and load distribution from the vehicle to the air-cushion system. A fuzzy logic expert system (FLES) will be introduced in this chapter to maintain the cushion pressure at any instant of the vehicle position during traversing over the terrain. The entire methodology that are involved on the development of intelligent air-cushion tracked vehicle is shown in Figure 8.1.

Assumptions of the Development of Intelligent Air-cushion Tracked Vehicle

The tractive performance justification of an off-road tracked vehicle is so important that it ensures the vehicle mobility over the unprepared peat terrain. The vehicle tractive performance for the unprepared peat terrain was determined with accounting the vehicle weight, track ground contact length, width, pitch and grouser height, vehicle ground pressure distribution, sinkage, slippage and trim angle, track entry and exit angle, sprocket and idler location and diameter, road-wheel number and diameter, geometrical arrangement and spacing, supporting roller diameter and geometrical arrangement, center of gravity location and height and track initial tension.

The tractive performance on the field was justified with the vehicle ground contact pressure distribution, tractive effort, motion resistance, and tractive efficiency. In this study, the well defined tracked vehicle performance over the swamp peat terrain was justified by using the fuzzy logic controlled air-cushion system.

The performance analysis was inferred by a study of vehicle design features such as: track sinkage, ground pressure distribution, track size, sprocket diameter, road-wheel diameter, number of road-wheel, center of gravity location, grouser height and traveling speed, As for simplification, there are several assumptions are made in order to study the vehicle design parameters. The assumptions are:

1. The pressure distribution in the track-terrain boundary is assumed to be identical by positioning vehicle C.G at the middle of the track system.
2. In developing the mathematical model for the vehicle, the track is assumed to be a medium pitch rigid link track.
3. The ground pressure distribution of the vehicle over the entire track ground contact length from front idler to rear sprocket is considered to be same.
4. Allowable sinkage of the vehicle is considered to be 70 mm based on the study of Jamaluddin (2002).
5. The track segment between two road-wheels is assumed to take up no load because of zero track deflection based on Wong (1998).
6. Air-cushion system initial height is considered to be 135 mm based on Luo et al. (2003).

7. An optimal load distribution ratio (δ) is considered to be 0.2 based on Luo et al. (2007).
8. The motion resistances from aerodynamics and track belly drag component in the computations of the total motion resistance are not significant based on Wong (2008).

The novel-design-air-cushion supporting (protecting) system made of aluminum alloy is shown in Figure 8.2. This system is used to protect the air-cushion system from the external threat on the ground by adjusting its vertical and longitudinal displacements automatically.

The air-cushion system is attached with the air-cushion tracked vehicle (ACTV) only for preventing the vehicle sinkage and increasing the vehicle floatation capacity.

It would be incurred once the vehicle transfers its load to the air-cushion system. Load transferring from the vehicle to the air-cushion system starts when the vehicle sinkage is 70 mm or more.

Figure 8.3 shows the vehicle tracked system which is divided into the four sections: the section (S1) in the upper part of the track supported by the supporting wheels; the section (S2) in contact with the front guider and the end of the first road-wheel; the section (S3) in contact with the road-wheels; and the section (S4) is wrapped with the rear sprocket.

The section of the track of front guider and the end of the first road-wheel is set at 20° in order to overcome the front obstacles if any and climb the steep terrains. The flow chart of the simulating model is shown in Figure 8.4.

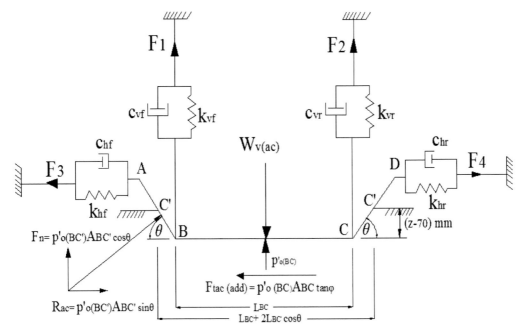

Source: Rahman et al. 2010.

Figure 8.2. Novel-design of air-cushion support system.

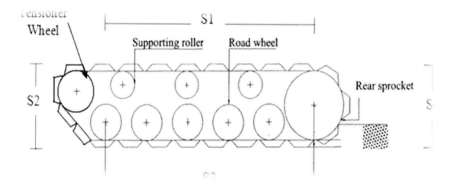

Figure 8.3. Track system for the swamp peat terrain vehicle.

Consider a track vehicle of total weight W as shown in Figure 3.3, track size including track ground contact length L, track width B, pitch T_p, and grouser height H, radius of the rear sprocket R_{rs}, and radius of the road-wheels R_r, and height of the center of gravity h_{cg} is traversing under traction on a swamp peat terrain at a constant speed of v_t as soon as applying the driving torque Q at the rear sprocket. The pressure distribution in the track-terrain boundary is assumed to be identical by positioning vehicle C.G at the middle of the track system.

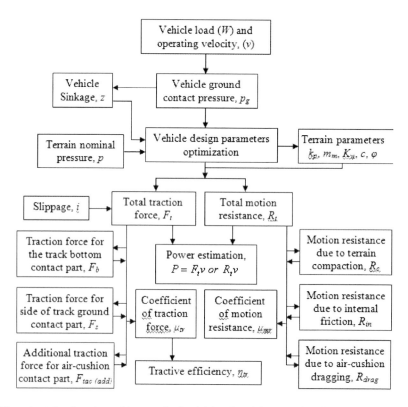

Figure 8.4. Flowchart for the air-cushion tracked vehicle. Consider a track vehicle of total weight W as shown in Figure.

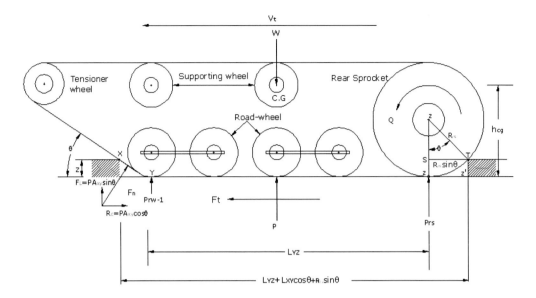

Figure 8.5. Forces acting on the vehicle track system.

The terrain pressure at main straight part p, and rear sprocket p_{rs}, and vehicle sinkage z are exposed same as C.G of the vehicle is positioned at the mid-point of the track system.

8.2.1. Sinkage

On soft terrain vehicle sinkage significantly affects on the vehicle total performance. It is noted that if the sinkage of the vehicle is more than or equals to the vehicle critical sinkage (i.e., $z = 70$ mm), the vehicle will stuck. Since the vehicle CG is located at the middle of the wheel base, the sinkage of the vehicle is computed based on the main straight part of the track system. Section S3 shown in Figure 8.3 is considered as the main straight part of the track system. The sinkage of the main straight part of the track system is as follows (Rahman et al., 2010a):

1. Vehicle track system without air-cushion:

$$z_1 = \frac{-\left(\frac{k_p D_{ht}}{4m_m}\right) \pm \sqrt{\left[\left(\frac{k_p D_{ht}}{4m_m}\right)^2 + \frac{D_{ht}}{m_m} p_g\right]}}{2} \quad (8.1)$$

where $D_{ht} = \dfrac{4LB}{2(L+B)}$, $L = L_{YZ}$, $p_g = \dfrac{W}{A_t}$, where, $A_t = 2(L \times B)$

2. Vehicle track system with air-cushion:

$$z_2 = \frac{-\left(\dfrac{k_p D_{htc}}{4m_m}\right) \pm \sqrt{\left[\left(\dfrac{k_p D_{htc}}{4m_m}\right)^2 + \dfrac{D_{htc}}{m_m}p_g\right]}}{2}$$

(8.2)

where

$$D_{htc} = \frac{4BL_{tc}}{2(L_{tc}+B)}, \quad L_{tc} = \left(L_{XY}\cos\theta + L_{YZ} + R_{rs}\sin\theta\right) L_{XY} = \frac{z}{\sin\theta},$$

$$p_g = \frac{W_t}{A_{tc}},$$

where $A_{tc} = 2(L_{tc} \times B)$ and $W_t = W - W_{v(ac)} = (1-\delta)W$

In equation (3.3-3.4), p_g is the vehicle ground contact pressure in kN/m^2 and z is the sinkage in m, m_m is the surface mat stiffness in kN/m^3, k_p is the underlying peat stiffness in kN/m^3, D_{ht} is the track hydraulic diameter in m, D_{htc} is the track hydraulic diameter in m when air-cushion touches the ground, B is the track width in m, L is the track ground contact length in m (L_{YZ} in Figure 3.3), L_{tc} is the ground contact length of the track in m when air-cushion touches the ground, A_t is the track ground contact area in m^2, A_{tc} is the track ground contact area in m^2 when air-cushion touches the ground, W is the total vehicle load in kN, W_t is the vehicle load supported by the track system in kN, $W_{v(ac)}$ is the vehicle partial load supported by the air-cushion system in kN, δ is the load distribution ratio, R_{rs} is the radius of rear sprocket in m, and θ is the angle between the track of the 1st road-wheel to tensioned wheel and to the ground in degrees as shown in Figure 3.3. If $W_{v(ac)} = 0$, i.e., when the vehicle sinkage is less than 70 mm, then air-cushion is not inflated and therefore, $W_t = W$.

8.2.2. Traction Force

Traction force (or tractive effort) of a track developed by shearing of the terrain determines the performance potential of the vehicle. A tracked vehicle develops traction force by deforming the soil in longitudinal shear. As the vehicle moves across the soil, a counter force arises from the soil and is equal to the traction force. In other sense, it can be defined as the force of the vehicle that can be mobilized at the terrain interface. The traction force is developed not only at the ground contact part of the track (F_b) but also at the track sides (F_s) due to the shearing action on the vertical surfaces on either side of the track with high grousers. The traction force equation for the vehicle's track ground contact part on swamp peat terrain is modeled by the shearing of the terrain and is computed by using the general equation of Bekker (1969) as:

$$F = 2B \int_0^L \tau \, dx \tag{8.3}$$

In Eq. (8.3), F is the traction force (or tractive effort) in kN, τ is the shearing strength in kN/m^2 and x is the longitudinal distance in m. The shearing strength τ can be computed as a function of the shear displacement and of the normal pressure by using Wong (2001) as:

$$\tau = \tau_{max} \left(j / K_w \right) \exp \left(1 - j / K_w \right)$$

where j is the shear displacement in m, and K_w is the shear deformation modulus in m where the maximum shear stress τ_{max} occurs and it can be expressed as

$$\tau_{max} = c + p_g \tan \varphi = c + \frac{W_t}{A_t} \tan \varphi \tag{8.4}$$

where c is the cohesiveness in kN/m^2, φ is the terrain internal friction angle in degrees, and p_g is the vehicle ground contact pressure in kN/m^2. The shear displacement j at a point located at a distance x from the front of the contact area can be determined by

$$j = v_j t$$

where v_j is the slip velocity of the vehicle track with reference to the ground in m/s, t is the travelling time over j in sec and is equal to $\dfrac{x}{v_t}$, where v_t is the theoretical velocity of the vehicle in m/s. Rearranging the above-mentioned equation, the expression for shear displacement j becomes

$$j = \frac{v_j x}{v_t} = ix$$

where i is the slippage (or slip ratio) of the vehicle in percentage.

Combining the above-mentioned equations into equations (8.2) and (8.4), traction force for the vehicle's track ground contact part on swamp peat terrain can be calculated as follows:

$$F_b = 2B \int_0^L \left(c + \frac{W_t}{A_t} \tan \varphi \right) \left(\frac{ix}{K_w} \right) e^{\left(1 - \frac{ix}{K_w} \right)} dx \tag{8.5}$$

For simplifying the equation (8.5), $z = 1 - \dfrac{ix}{K_w}$ is assumed. Thus, the equation (8.5) can be written as

$$F_b = 2B\left(c + \frac{W_t}{A_t}\tan\varphi\right)\int^{\left(1-\frac{iL}{K_W}\right)}(1-z)e^z\left(-\frac{K_w}{i}\right)dz$$

$$= (A_t c + W_t\tan\varphi)\left[\frac{K_w}{iL}e^1 - \left(1 + \frac{K_w}{iL}\right)e^{1-\frac{iL}{K_W}}\right]$$

(8.6)

The traction force for the vehicle's track ground contact part on swamp peat terrain is calculated in this study for the two sinkage conditions:

1. For sinkage, $0 \le z \le 70$ mm (Vehicle without air-cushion):

$$F_b = (A_t c + W_t\tan\varphi)\left[\frac{K_w}{iL}e^1 - \left(1 + \frac{K_w}{iL}\right)\exp\left(1 - \frac{iL}{K_w}\right)\right]$$

(8.7)

where $A_t = 2(L)(B)$ and $L = L_{YZ}$

1.   For sinkage, $z \ge 70$ mm (Vehicle with air-cushion):

$$F_{bc} = (A_{tc} c + W_t\tan\varphi)\left[\frac{K_w}{iL_{tc}}e^1 - \left(1 + \frac{K_w}{iL_{tc}}\right)\exp\left(1 - \frac{iL_{tc}}{K_w}\right)\right] + F_{tac\,(add)}$$

(8.8)

where $A_{tc} = 2(L_{tc})(B)$, $L_{tc} = (L_{XY}\cos\theta + L_{YZ} + R_{rs}\sin\theta)$,

and

$$L_{XY} = \frac{z}{\sin\theta}$$

In equation (8.5-8.8), F_b is the traction force developed at the vehicle's track ground contact part in kN, F_{bc} is the traction force developed at the vehicle's track ground contact part in kN when air-cushion touches the ground, $F_{tac\,(add)}$ is the additional traction force which is developed due to the sliding of the air-cushion system over the terrain in kN, B is the track width in m, L is the track ground contact length in m, L_{tc} is the track ground contact length in m when air-cushion touches the ground, A_t is the track ground contact area in m^2, A_{tc} is the track ground contact area in m^2 when air-cushion touches the ground, W_t is the vehicle load supported by the track system in kN, c is the cohesiveness in kN/m^2, φ is the terrain internal friction angle in degrees, K_w is the shear deformation modulus of the terrain in m, i is the slippage of the vehicle in percentage, e is the exponent (exp) term, z is the sinkage of the vehicle in m, R_{rs} is the radius of rear sprocket in m, and θ is the angle between the track of the 1st road-wheel to tensioned wheel and to the ground in degrees.

The traction mechanics of the track at the side of the grouser is highly significant on the development of vehicle traction if the vehicle sinkage is more than the grouser height (Wong, 2001).

In this study, it is assumed that the sinkage of the vehicle is more than the grouser height of the track. For a track with high grousers, additional traction force is developed due to shearing action on the vertical surfaces on either side of the track. This additional traction force developed at the side of the ground contact length of the track can be represented by simplifying the recommended equation of Reece (1965-66), Wong (2001) and Ataur et al. (2006a) given first in general formulation:

$$F_s = 4H \cos \alpha \int_0^L \tau dx$$

The shear stress τ can be computed as a function of the shear displacement and of the normal pressure as:

$$\tau = \tau_{max} \left(j / K_w \right) \exp \left(1 - j / K_w \right)$$

where $\tau_{max} = c + p \tan \varphi$ and $j = ix$

Assuming a uniform normal pressure distribution, the traction force developed at the side of the vehicle's track, F_s has been finally computed as:

$$F_s = 4HL \left(c + p_g \tan \varphi \right) \cos \alpha \left[\frac{K_w}{iL} e^1 - \left(1 + \frac{K_w}{iL} \right) \exp \left(1 - \frac{iL}{K_w} \right) \right] \tag{8.9}$$

The traction force developed at the side of the vehicle's track on swamp peat terrain is calculated in this study for the two sinkage conditions:

1. For sinkage, $0 \leq z \leq 70$ *mm* (Vehicle without air-cushion):

$$F_s = 4HL \left(c + p_g \tan \varphi \right) \cos \alpha \left[\frac{K_w}{iL} e^1 - \left(1 + \frac{K_w}{iL} \right) \exp \left(1 - \frac{iL}{K_w} \right) \right] \tag{8.10}$$

2. For sinkage, $z \geq 70 \ mm$ (Vehicle with air-cushion):

$$F_{sc} = 4HL_{tc}\left(c + p_g \tan \varphi\right)\cos \alpha \left[\frac{K_w}{iL_{tc}}e^1 - \left(1 + \frac{K}{iL_{tc}}\right)\exp\left(1 - \frac{iL_{tc}}{K_w}\right)\right] \qquad (8.11)$$

where $L_{tc} = \left(L_{XY}\cos\theta + L_{YZ} + R_{rs}\sin\theta\right), L_{XY} = \dfrac{z}{\sin\theta}, p_g = \dfrac{W}{A_{tc}}$ and $A_{tc} = 2\left(L_{tc}\right)\left(B\right)$

In Eq. (8.10-8.11), F_s is the traction force developed at the side of the tracks in kN, F_{sc} is the traction force developed at the side of the tracks in kN when air-cushion touches the ground, H is the grouser height in m, p_g is the vehicle ground contact pressure in kN/m^2, B is the track width in m, L is the track ground contact length in m, L_{tc} is the track ground contact length in m when air-cushion touches the ground, A_t is the track ground contact area in m^2, A_{tc} is the track ground contact area in m^2 when air-cushion touches the ground, c is the cohesiveness in kN/m^2, φ is the terrain internal friction angle in degrees, K_w is the shear deformation modulus of the terrain in m, i is the slippage of the vehicle in percentage, e is the exponent (exp) term, α is the slip angle in degree, z is the sinkage of the vehicle in m, R_{rs} is the radius of rear sprocket in m, and θ is the angle between the track of the 1st road-wheel to tensioned wheel and to the ground in degrees

8.2.3. Motion Resistance

The motion resistance of a tracked vehicle is mainly incurred due to the terrain compaction which is referred as a terrain compaction motion resistance, R_c. The motion resistance R_c can be modeled by considering the work done with towing the vehicle a distance L in the horizontal direction.

$$R_c L = U$$

The work done in compressing the terrain and creating a rut of width B, length L, and depth z is given by

$$U = 2BL \int_0^z pdz$$

$$R_c = 2B\left(\frac{k_p z^2}{2} + \frac{4}{3D_h}m_m z^3\right) \qquad (8.12)$$

where R_c is the vehicle motion resistance due to terrain compaction to two tracks in kN. In this study, the motion resistance of the vehicle due to terrain compaction is computed by considering:

1. $0 \le z \le 70\ mm$ and (ii) $z > 70\ mm$.
2. For sinkage, $0 \le z \le 70\ mm$

$$R_c = 2B \left(\frac{k_p z^2}{2} + \frac{4}{3 D_h} m_m z^3 \right) \qquad (8.13)$$

where $z = \dfrac{-\left(\dfrac{k_p D_{ht}}{4 m_m} \right) \pm \sqrt{\left[\left(\dfrac{k_p D_{ht}}{4 m_m} \right)^2 + \dfrac{D_{ht}}{m_m} p_g \right]}}{2}$,

$$D_{ht} = \frac{4 BL}{2(L + B)} \text{ and } p_g = \frac{W_t}{A_t}$$

3. For sinkage, $z > 70\ mm$

$$R_{c'} = 2B \left(\frac{k_p z^2}{2} + \frac{4}{3 D_{htc}} m_m z^3 \right) \qquad (8.14)$$

where $R_{c'}$ is the vehicle motion resistance due to terrain compaction to two tracks in kN when air-cushion touches the ground. The drag motion resistance R_{drag} of the air-cushion system based on the Figure 8.2 could be computed as follows:

1. For sinkage, $z = 0.0\ mm$

$$R_{drag} = 0 \qquad (8.15)$$

2. For sinkage, $z = 70\ mm$

$$R_{drag} = p_c A_c \tan \varphi = \left(p_g - p \right)\left(A_c \right) \tan \varphi \qquad (8.16)$$

3. For vehicle sinkage, $z > 70\ mm$

$$R_{drag} = p_c \left(A_{BC'} + A_{CC'} \right) \sin \theta + p_c A_c \tan \varphi \qquad (8.17)$$

where p_c is the pressure exerted on the air-cushion system due to the vehicle's dynamic load transfer to the air-cushion system in kN/m², A_c is the contact area of the air-cushion support system in m², and φ is the terrain internal friction angle in degrees.

The motion resistance of the vehicle due to internal friction losses, deeply affected by the track and the speed of the vehicle, is computed by the following empirical equation mentioned by Wong (2001):

$$R_{in} = \left(\frac{W_t}{1000\ g}\right)[222 + 3v] = \left(\frac{W - W_c}{1000\ g}\right)[222 + 3v] \quad (8.18)$$

where R_{in} is the motion resistance of the vehicle for internal friction of the moving parts in kN, W_t is the load supported by the track system in kN, W is the total vehicle load in kN, W_c is the load supported by the air-cushion in kN, g is the gravitational acceleration in m/s², and v is the velocity of the vehicle in km/h.

8.3. INTELLIGENT AIR-CUSHION SYSTEM

A cushion pressure control system with fuzzy logic controller will be designed to realize the cushion pressure targets and thus minimize the total power consumption based on the sinkage. The controller can overcome the disadvantage of the conventional PID with unadjustable parameter setting. It has the advantage of fuzzy controller being simple (relations between input and output variables can be explained in a linguistic-based rule base), robust (performance is not depending on training and new input variables and rules can be easily added) and not requiring precise mathematical model (Wang et al., 1999; Carman, 2008).

In the control system of the vehicle, position (vehicle vertical distance) h will be selected as controlled variable, and air flow rate Q as regulated variable through the change in valve position. Figure 8.5 will be the model of control system for the intelligent air-cushion system.

In Figure 8.5, the measured cushion position (h) and reference cushion position (hr), the position is controlled by a regulation variable, i.e., flow rate Q. The reference position hr is calculated based on the maximum allowable sinkage and then will be compared with the measured position values using distance sensor.

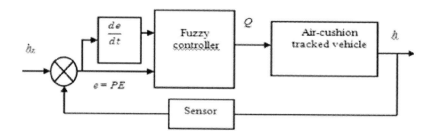

Figure 8.5. Block diagram of the control system.

Hence, the resultant deviation, i.e., position error (PE), e and differential position or rate of position error (RPE) $\dot{e}$ will continuously measured in operation. From the dynamics of the system, if the actual position will be less than the desired position, it will be necessary to increase the vehicle floatation by regulating flow rate through the change in inlet valve position to inflate the air-cushion and vice versa. This knowledge of the system behavior allows formulating a set of general rules that will described in the following section.

It is noted that the air-cushion will be started to inflate with decreasing the vehicle position. So the vehicle load will be slightly increased due to the air-cushion inflation pressure. Fuzzy controller used fuzzy logic expert system (FLES) will be introduced in this study for controlling intelligent air-cushion tracked vehicle position (sinkage) based on the discussion of (Passino and Yurkovich, 1998; Gopal, 2009; Hossain et al., 2010a). The intelligent air-cushion system is developed based on the discussion of section 3.5.1.

The interactions between the ground surface, instantaneous air supply to the cushion system, and their effects on the proposed vehicle dynamics are quite complex. The air-cushion serves as the primary coupling mechanism between the ground surface and the vehicle. The air-cushion dynamics is maintained with controlling air supply to the cushion based on the output signal of the distance measuring sensor. The air-cushion attached with the vehicle chassis frame has been approximated as a rectangular box for this study. The air-cushion model is based on adiabatic gas law and gives a static relationship between pressure and volume such that

$$pV^{\gamma} = const$$

$$p_i V_i^{\gamma} = p_f V_f^{\gamma} \qquad (8.19)$$

Considering an air-cushion cavity with constant length L_c (L_{BC} as shown in Figure 8.2) and width B_c, Eq. (8.19) can be written as (Hossain et al., 2010d):

$$p_i h_i^{\gamma} = p_f h_f^{\gamma}, \text{ Here volume, } V = \oint(h) \text{ as area, } A_{BC} = constant \qquad (8.20)$$

where p_i is the initial pressure in N/m², V_i is the initial volume of air-cushion in m³, h_i is the initial height of air-cushion in m, p_f is the final (inflated) pressure in the cushion in N/m², and h_f is the final height after air-cushion inflation in m.

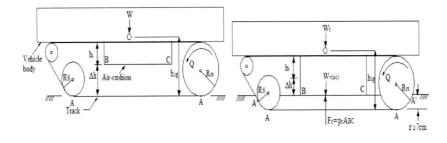

Figure 8.6. Depiction of the air-cushion inflation system.

Letting $h_f = h_i + dh$, where, dh is the change of cushion height:

$$p_i h_i^\gamma = p_f h_i^\gamma \left(1 + \frac{dh}{h_i}\right)^\gamma \qquad (8.20)$$

with $p_f - p_i = \dfrac{-p_f}{h_i}\gamma dh$

where negative sign indicates the upward direction (since downward direction is assumed as positive) of the resultant pressure between final (full) inflated pressure and initial inflated pressure.

The use of mathematical models with control parameters in air-cushion system for tracked vehicle operating on swamp terrain which involve various types of uncertainties and vague phenomena raises the problems how accurately they reflect reality. Hence it is natural to look for different methodologies. In this regard, the fuzzy logic expert system (FLES) for automotive engineering is extended to vehicle dynamics, in particular to an air-cushion-terrain system. The air-cushion-terrain interaction takes place in an uncertain and vague environment due to soil conditions, submerged and undecomposed materials, interaction with other environments not accounted in the system, etc. No mathematical model can describe satisfactory such a difficult system. The theoretical models can be expected to drive performance rules of wide-ranging nature about the interacting cushion-terrain system but all are bound to make simplifying assumptions. The control objective of the cushion pressure system management is to regulate volume flow rate through the change in valve position by using a fuzzy logic expert system. Figure 8.7 illustrates the basic scheme for vehicle vertical position control (i.e., sinkage control) during sinking due to the low bearing capacity swamp peat terrain. In this figure, two valves control the inlet and outlet flow rates of the air pressure, respectively. A distance sensor is mounted at the vehicle chassis frame to measure the vehicle vertical position from which the vehicle sinkage is calculated.

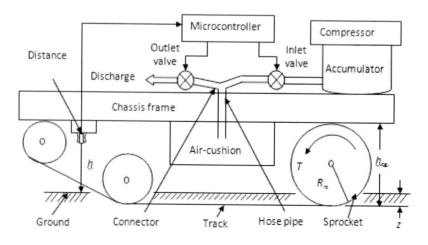

Figure 8.7. Vehicle vertical position control problem.

8.4. STUDY OF LOAD DISTRIBUTION RATIO

The effects of slippage on load distribution ratio for the intelligent air-cushion tracked vehicle (IACTV) have been investigated. It is observed that slippage is decreasing significantly with the increase of load distribution ratio (defined as the ratio of vehicle load supported by the cushion system to the total vehicle load). In order to maintain the air-cushion tracked vehicle in normal operating conditions, i.e. with satisfactory performance, slippage i and load distribution ratio δ should be respectively kept within the range of 7-30% and 0.12-0.45. The data are presented in Figure 8.8, demonstrating that the load distribution ratio around 0.26 gives optimum value of slippage of 15% which could be supported by Luo et al. (2003). Moreover, the relationship between load distribution ratio and slippage might be useful for sufficient traction control for the vehicle normal operation, braking and positioning.

For a given terrain condition, the required traction (driving) force of the vehicle can be predicted with different slippage and different load distribution levels. It is observed that the load distribution ratio and total resistances decrease significantly with the increase of vehicle slippage. Consequently, the traction force steeply increases to a maximum value with the increase in slippage and again linearly decreases with the further increase of slippage.

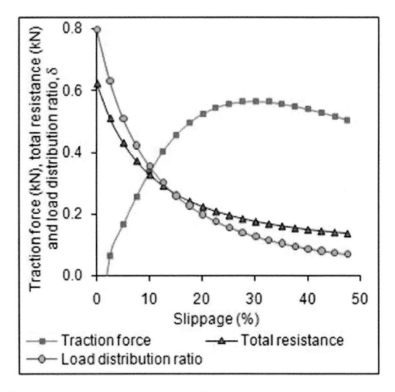

Figure 8.8. Effects of force, resistance and load distribution ratio on vehicle slippage.

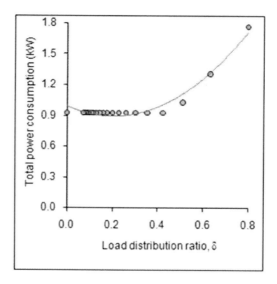

Figure 8.9. Effects of load distribution ratio on total power consumption.

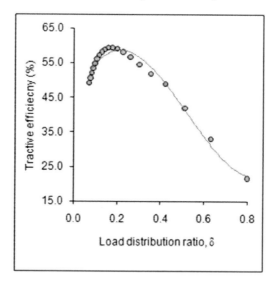

Figure 8.10. Effects of load distribution ratio on tractive efficiency.

Accordingly, the required traction force increases rapidly with the decreases of load distribution ratio up to a maximum value and then decreases slowly with further increase of load distribution ratio. However, it can be noticed that the vehicle slippage should be reasonably maintained within the range of 7-30 % in order to overcome the travelling resistance and maintain normal driving condition to ensure normal operations. Since the current study is focusing on load distribution control system intending at reducing total power consumption, the effect of load distribution on the total power consumption is also examined. According to previous studies (Shuo et al., 2006), it has been shown that load distribution ratio affects total power consumption significantly. If the vehicle with a constant load is affected by external disturbances such as wood, shrimps, slugs, stones, etc., the cushion clearance height of the vehicle and air-cushion pressure will change, and hence the required power for the vehicle will also change. Consequently, this results in the changes of load distribution ratio and total power consumption. Figure 8.9 shows the relationship between load distribution ratio δ and total power consumption.

It is noticed that the load distribution ratio affects the total power consumption extensively as it linearly increases with the increase of load distribution ratio. For a particular terrain situation, an optimal load distribution ratio exists which guide to a minimum power consumption. However, for the present study, an optimal load distribution ratio of 0.26 is obtained which results in minimum total power consumption of 0.930 kW for the vehicle load of 1.96 kN. Furthermore, when load distribution ratio exceeds about 0.4, the total power consumption increases significantly. Obviously, in different operating conditions, there is minimal theoretical power consumption with respect to i and δ.

Furthermore, tractive efficiency is an important criterion to evaluate the trafficability of the vehicle. Figure 8.10 shows the relationship between the variations of load distribution ratio with the variation of tractive efficiency varied from 20 to 60%. It is noticed that as load distribution ratio δ increases from 0, tractive efficiency increases rapidly and the curve reaches its peak; then gradually decreases with the further increase of δ. Based on the established theoretical model and the developed prototype, corresponding simulation and experimental results have been conducted and an optimal load distribution ratio of 0.26 has been attained based on previous studies (Luo and Yu, 2007) which could result in prediction of maximum tractive efficiency of 56%. However, in starting acceleration case, the actual load distribution ratio needs to be controlled in order to get maximal tractive efficiency. When the vehicle is in uniform velocity motion, driving force equals travelling resistance, so tractive efficiency could be equal to zero. Based on the Figure, it is concluded that cushion system will be stuck on the terrain if the load distribution increases more than 50%. Hence, the vehicle propulsion system is unable to propel the vehicle.

In practice, some uncertainties inheritably exist, e.g., terrain surface conditions and vehicle loading conditions can be changeable. Therefore, it should be noted that the present study was only focused on vertical load distribution control using air-cushion system for avoiding sinking during normal driving cases, implying the slippage within an acceptable range. Beyond this slippage limitation range, e.g., exceeding about 60%, the theoretical model developed may not be valid.

In the earlier study it is reported that the bearing capacity of the swamp peat can be the terrain nominal pressure (TNP). Figure 8.11 shows the significance of the air-cushion system for the vehicle potentiality over the swamp peat terrain where TNP is lower than 7 kN/m^2. Vehicle ground contact pressure for the 1.96 kN and 2.45 kN respectively are more than the terrain nominal pressure as shown in Figure 8.11(a) and 8.11(b). Therefore, the vehicle would not traverse over the swamp terrain. By equipping air-cushion system of contact area 0.144 m^2, the vehicle is fit to traverse on the terrain. Terrain nominal pressure p is defined as the pressure that exists from the ground with respect to the static or dynamic load of the vehicle on that ground.

Figure 8.12 shows the load distribution to air-cushion is all over the vehicle's travelling distance for the two loading conditions as stated before. It is seen that the power consumption of the vehicle depends on the load distribution to the air-cushion system.

It is observed that the maximum power requirement of the vehicle: 1.28 kW for the 1.96 kN vehicle and 2.2 kW for the 2.45 kN vehicle respectively. Since the vehicle is powered by the battery pack, emphasis has been given to less load distribution to the air-cushion system which is operated with the developed additional thrust and the compressor motor consumes 50-64% of the total power of the battery pack.

From the simulation studies and previous analysis, the vehicle's total ground contact area is considered as 0.282 m^2 including 0.144 m^2 of air-cushion effective area has been optimized based on the bearing capacity of the terrain. Consequently, the vehicle tracked ground contact area 0.138 m^2 is optimized for working on the moderate peat terrain without activating the air-cushion system.

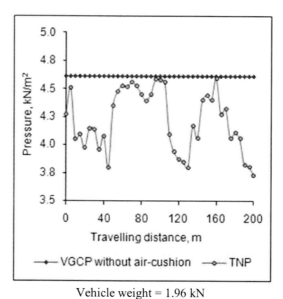

Vehicle weight = 1.96 kN

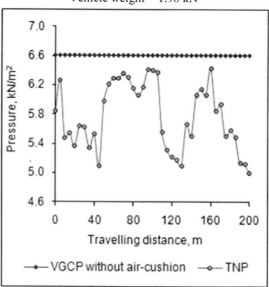

Vehicle weight = 2.45 kN

Figure 8.11. Variation of the vehicle ground contact pressure (VGCP) and terrain nominal pressure (TNP) for a travelling distance of 200 m.

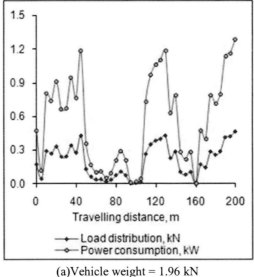

(a) Vehicle weight = 1.96 kN

(b) Vehicle weight = 2.45 kN

Figure 8.13. Variation of the load distribution and total power consumption.

Figure 8.14 shows the relationship between the traction force and motion resistance for the vehicle weight of 1.96 kN.

The result demonstrates that the vehicle is able to travel the terrain as the vehicle total traction force (Ft) is more than the vehicle's total motion resistance (Rt) which is mainly due to the terrain compaction and air-cushion system dragging. The track system is unable to develop sufficient traction force (Fb) to overcome the motion resistance and this could be for the lower cohesiveness (c) and internal friction angle (φ). For the vehicle ground contact area of 0.282 m^2, total motion resistance and total traction force are found as 1.03 kN and 1.33 kN, respectively. It is noticed that traction force developed by the tracked system is 0.69 kN while the additional 0.64 kN force is developed by the propeller. Therefore, it could be concluded that the vehicle can travel over the terrain with the aid of air-cushion system equipped with a propeller.

Figure 8.15 shows that the vehicle needs to develop total power of 4.05 kN to overcome the total motion resistance (Rt) for the vehicle ground contact area of 0.282 m^2. From the figure, it is oticed that the motors need 2.09 kW power to build up the adequate torque to the driving sprockets so that it can develop 0.69 kN traction force to the tracked system.

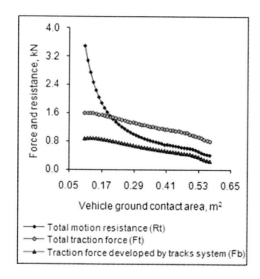

Figure 8.14. Relationship between the traction force and motion resistance.

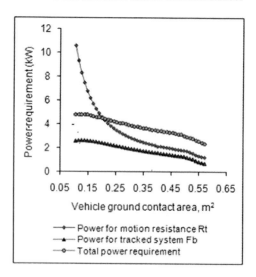

Figure 8.15. Power requirement for propelling the vehicle.

Meanwhile, the propeller requires an additional power of 1.96 kW to develop the 0.64 kN additional thrust for the vehicle to travel the terrain. Therefore, the vehicle requires selecting a battery pack which is able to supply 4.05 kW for 2 h during traveling the swamp peat terrain. Eight (8) batteries are considered in the battery pack in order to control the motor of the tracked system and propeller motor. It is noted that the vehicle is considered primarily activated for 2 h with the power of the battery pack.

8.5 DESCRIPTION OF INTELLIGENT AIR-CUSHION VEHICLE DEVELOPMENT

Figure 8.16 shows the developed air-cushion tracked vehicle with a moderate weight of 1.96 kN including 0.981 kN payload. The total ground contact area of the vehicle is 0.282 m^2 including 0.138 m^2 of track ground contact area and 0.144 m^2 of air-cushion area which would provide the vehicle lower ground contact pressure of 6.96 kN/m^2. This would allow the vehicle lower sinkage and rolling resistance, thus providing the high tractive effort yielding higher level of tractive performance.

Figure 8.16. Developed air-cushion tracked vehicle.

Steering of this vehicle is achieved by means of an individual switch of the DC motor with a power of 0.500kW@2.94 Nm. The vehicle is powered by a battery pack comprising eight (8) lead acid batteries. The vehicle can travel 1 hour powered of the single charging battery pack. A small IC Engine power of 2.5 kW@4000 rpm is mounted on the vehicle to recharge the battery pack with the aid of an alternator. The vehicle is controlled by a remote system from the side of road in order to keep away from the risk if there is any possibility. General view of the air-cushion tracked vehicle has been shown in Figure 8.17 (Rahman et al., 2010b).

The air-cushion tracked vehicle is included mainly two systems: the track (propulsion) system which consists of two full segmented rubber tracks mechanism as driving system to overcome traveling resistance, and the air-cushion system as vehicle body to increase the floatation capacity of the vehicle. The vehicle components are comprised mainly with the full track system, air-cushion system, two DC motors, a propeller, battery pack, and a small engine. The vehicle power transmission system is shown in Figure 8.18.

Notation: 1-Batter pack, 2-Engine, 3-Alternator, 4-Drive pulley, 5-Engine's pulley, 6-Rear sprocket, 7-Track cover, 8-Road wheel

Figure 8.17. General view of air-cushion tracked vehicle.

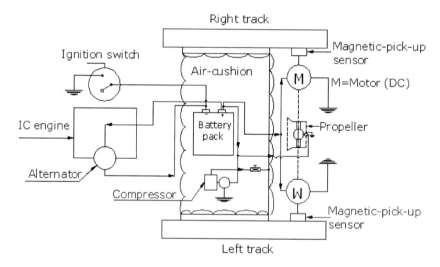

Figure 8.18. Vehicle power transmission system.

The driving force is provided to each of the tracks system by an individual DC motor. The air-cushion system is mounted with the inner sides of the track system and just 70 mm from the bottom of the track system. The additional thrust is provided to the air-cushion system by a propeller. The vehicle could run on the unprepared swamp peat terrain at least 100 km from a single charging battery. After traveling 90 km the battery is recharged by the engine automatically. The automatic recharging system of the battery is incorporated by using a sensor. The air-cushion system makes the vehicle ground contact pressure less than 7 kN/m2.

The hovering pressure to the air-cushion is provided by a single compressor to lift the vehicle. The starting system of the compressor is incorporated by using a distance sensor. The DC motors, the compressor, and the propeller receive power from a rechargeable high amp-hr battery.

Figure8.19. Intelligent system on air-cushion tracked vehicle.

8.6. VEHICLE PERFORMANCE INVESTIGATION

To investigate the vehicle performance in terms traction force (tractive effort) and motion resistance, the experiment of air-cushion tracked vehicle has been performed. The experimental set-up consists of tracks system for propulsion and air-cushion system for lifting the vehicle with two kinds of experiments: laboratory and field experiment. The vehicle is tested for both cases under controlled condition of the cohesiveness of the field considered as approximately constant for all the travelling length. The vehicle performance investigation has been conducted without and with intelligent system.

8.6.1. Laboratory Experiment without Intelligent System

The vehicle is tested outside on the field of the Automotive Laboratory, Faculty of Engineering, International Islamic University Malaysia (IIUM) at travelling speed of 10 km/h with loading conditions of 1.96 kN and 2.45 kN on normal surface with and without activating the air-cushion system. The vehicle travelling distance was considered as 50 m during testing.

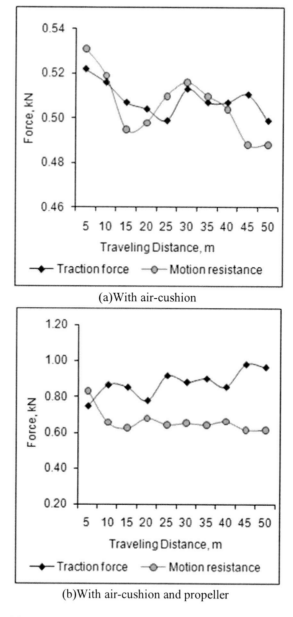

(a) With air-cushion

(b) With air-cushion and propeller

Figure 8.20. Variation of force and motion resistance for the vehicle weight 1.96 kN.

The output torque of the DC motor was measured by using a digital multimeter and it was converted into traction force. The motion resistance test was conducted by drawing the vehicle with an auxiliary vehicle. It is noted that a load cell was placed in between the tested and auxiliary vehicle. The vehicle was found to have much potential to travel over the grass field without the air-cushion. However, it was stuck once the air-cushion touches with the terrain. By using the propeller additional thrust it was activated without getting stuck. The tractive performance data were used to prepare a data sheet and the values of traction force and motion resistance were calculated. Figures 8.20 and 8.21 show the typical variation of traction force and motion resistance of the air-cushion tracked vehicle for the two loading conditions of 1.96 kN and 2.45 kN, respectively. The results demonstrate that the mean value of traction force with a propeller over the without a propeller increases 10.21% and 6.47% for the vehicle weights of 1.96 kN and 2.45 kN, respectively. Similarly, it was found that the motion resistance decreases 12.63% and 24.81% for the vehicle weights of 1.96 kN and 2.45 kN, respectively

8.6.2. Field Experiment without Intelligent System

The field experiment was performed on the terrain of length 50 m which was soft soil similar to swamp peat at the Faculty of Engineering, IIUM. The terrain used in testing was soft with small grass and little amount of water to make similar to swamp peat. Figures 8.22(a) and 8.22(b) show the typical variation of traction force of the air-cushion tracked vehicle considered with weight of 1.96 kN and 2.45 kN, respectively, for travelling speed of 10 km/h. It is noticed that the mean values of traction are 0.62 kN and 1.06 kN for the vehicle weight of 1.96 kN and 2.45 kN, respectively. Meanwhile, with increasing the vehicle loading conditions from 1.96 kN to 2.45 kN, the mean values of the vehicle traction increases 71%. This significant amount of traction increasing is principally due to the loading situation of the vehicle as the cohesiveness of the field is approximately invariable for the entire terrain length. Furthermore, this trend could be due to the hydrodynamic effect of the terrain as there was no drainage system in the field (Wong et al., 1982). However, from the experiments, it was noticed that the air-cushion tracked vehicle (Figure 8.17) was stuck once the air-cushion contacts with the terrain. By using the propeller's additional thrust the vehicle was operated without getting stuck. It appears that if the cushion system is in contact with the terrain all the while it needs more power to operate the propeller. It is also noticed that the air-cushion system is in contact with the terrain once it has been inflated. Therefore, in this study intelligent system is developed to operate the air-cushion system and to overcome the problem that has been incurred.

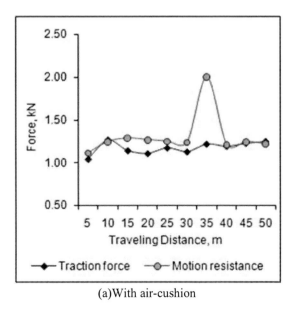

(a) With air-cushion

Figure 8.21. (Continued).

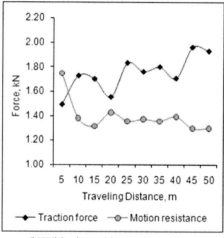

(b)With air-cushion and propeller

Figure 8.21. Variation of force and motion resistance for the vehicle weight 2.45 kN.

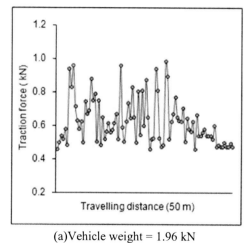

(a)Vehicle weight = 1.96 kN

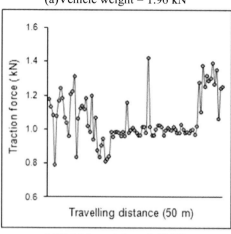

(b)Vehicle weight = 2.45 kN

Figure 8.22. Variation of traction force for a travelling distance of 50 m.

8.7. Vehicle Perfromance Prediction: Fuzzy Logic Approach

Earlier study shows that terrain conditions and air-cushion system significantly affect vehicle performance. To accurately predict the intelligent air-cushion tracked vehicle (IACTV) performance in terms of controlling air flow rate to the cushion chamber, traction force, motion resistance, tractive efficiency and total power consumption in a given soil and operating conditions, different techniques are known from the literature survey (Grisso et al., 2006; Luo and Yu, 2007; Park et al., 2008; Xie et al., 2009 and Hossain et al., 2010b). At present artificial intelligence system such as machine learning, neural network, genetic algorithms, etc., have largely been used in different areas including automotive industries. However, fuzzy logic expert system (FLES) might play an important role since it uses expert knowledge on controlling the particular system, it is flexible and it correctly estimates the unknown values of the modeled data, often improve performance and it has high level expression capability (Marakoglu and Carman, 2010). The choice of fuzzy set theory as the main analytical tooling is due to the good applicability of this approach to uncertain vehicle-terrain systems and dynamic processes (Ivanov et al., 2010 and Altab et al., 2010). Generally, fuzzy sets use the linguistic expressions instead of numerical values as compared to classical data sets.

For the prediction of the IACTV's tractive performance, the vehicle has been assumed to traverse with constant velocity on uneven terrain conditions. For implementation of fuzzy theory into the vehicle system, the fuzzy toolbox from MATLAB has been used. Fuzzy inference system (FIS) is the actual process of mapping from a given set of input variables to an output on a set of fuzzy rules. Four fundamental units such as fuzzification unit, the knowledge base (rule base), the inference engine and defuzzification unit are necessary for the successful application of fuzzy modeling approach. The details have been explained in the earlier Chapter. However, this section incorporates the construction of fuzzy knowledge-based model using *if-then* rules for the prediction of air flow rate through the valve opening, traction force, motion resistance, tractive efficiency and total power consumption based on Mamdani approach. Sampling data collected from the experiment have been used to validate the fuzzy models.

8.7.1. Prediction of Air Flow Rate

For information about the input variables in the fuzzy inference system "position", a distance (vehicle vertical position) measuring sensor is used. The sensor has been attached with vehicle chassis frame and the reference height (vehicle vertical position) has been set as 15 cm based on the allowable vehicle sinkage of 7 cm (Jamaluddin, 2002). Lyasko (2010c) has reported that a very important part of a track and wheel tractive performance study is the sinkage which is necessary for obtaining vehicle traction, motion resistance, etc. Therefore, it is desired that the vehicle position be maintained at a desired position in order to make sufficient traction control.

For prediction of air flows in air-cushion inflation by using fuzzy expert system, position error (PE) and rate of position error (RPE) are used as input parameters and air flow rate through valve opening (Q) is used as output parameter. The units of the used factors are: PE (cm), RPE (cm/s) and Q (%).

The results of the developed FLES have been compared with the experimental results. Sample data is shown in Appendix E. The mean of measured and predicted values (from FLES) on flow rate Q are 77.78 % and 70.29 %, respectively. The correlation between actual (measured) and predicted values (FLES) of flow rate in different working conditions have been shown in Figure 8.23.

The relationship is significant. The correlation coefficient of relationship has been found 0.971 which is significant in operation. Furthermore, the mean relative error of actual and predicted values from the FLES model on flow rate is found as 10.93 % which is almost equal to the acceptable limit of 10% (Carman, 2008). The goodness of fit of the prediction values from the FLES model is found as 0.91 which is close to 1.0 as expected.

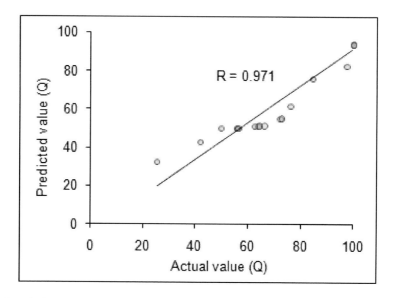

Figure 8.23. Correlation between actual and predicted values of flow rate.

8.7.2. Prediction of Vehicle Tractive Performance

The related mathematical model for the track-terrain interaction process has been used to investigate the vehicle tractive performance under a wide range of conditions with the aim to get better vehicle design and to optimize vehicle operational parameters. It is reported that the vehicle tractive performance could be measured by performing field testing which could be expensive (Tiwari et al., 2010). Because of this reason, fuzzy logic expert system (FLES) is used to predict the tractive performance in terms of traction force, motion resistance, tractive efficiency and power consumption for an intelligent air-cushion tracked vehicle. Predictions of all the parameters are mainly dependent on vehicle sinkage and weight. Lyasko (2010b) has reported that vehicle sinkage is a very important part of a track and wheel tractive performance study. He has mentioned that it is impossible to obtain vehicle traction force, motion resistance, soil trafficability, soil compaction, rut depth, etc. without it. In spite of the significant influence of the vehicle sinkage on motion resistance, slip sinakge should be considered. Reece (1965-66) has proposed an empirical formula to calculate vehicle sinkage as a sum of soil sinkage due to a normal load and additional sinkage due to a soil horizontal deformation (shear). However, Bekker (1969) has mentioned that the additional sinkage is originally small at low slippage and it reaches the height of the grouser at slip 50%. Since in this study, maximum slippage has been considered as 15 % and hence the sinkage has been calculated based on soil sinkage due to vehicle normal load only. Estimation of vehicle sinkage has been achieved through a distance measuring sensor that assesses the vehicle vertical position and hence measures the sinkage. Vehicle sinkage is considered in the range of 0.02-0.08 m while vehicle weight is in the range of 1.8-2.5 kN. It is noted that the integration of several physical characteristics as input variables is needed to improve robustness by recognizing the type of road surface, weather condition, etc. For this case the base of fuzzy rules can be significantly complicated, and the amendment of the output variables will be still small.

For implementation of fuzzy values into the system, vehicle sinkage (VS) and vehicle weight (VW) are used as input parameters and vehicle traction force (TF), motion resistance (MR), tractive efficiency (TE) and power consumption (PC) are used as output. For fuzzification of these factors the linguistic variables very low (VL), low (L), low medium (LM), medium (M), high medium (HM), high (H) and very high (VH) are used for the inputs and outputs. In this study, a Mamdani max-min inference approach and the center of gravity defuzzification method have been used because these operators assure a linear interpolation of the output between the rules (Carman, 2008). The Mamdani fuzzy inference system employs the individual rule based inference scheme, and derives the output subjected to a crisp input. Within the framework of the presented investigation, triangular shaped membership functions are used for both input and output variables because of their accuracy (Marakoglu and Carman, 2010).

Selection of the membership functions and their formations is based on the system knowledge, expert's appraisals, and experiment conditions (Ivanov et al., 2010). The units of the input and output variables are: VS (m), VW (kN), TF (kN), MR (kN), TE (%) and PC (kW). For the input and output parameters, a fuzzy associated memory is formed as regulation rules. Total of 49 rules have been formed. Parts of the developed rules are shown in Table 8.1. For example, rule 1 is interpreted as: If VS = VL and VW = VL, then TF = VH, MR = VL, TE = VH and PC = VL.

Table 8.1. Inference rules of prediction parameters

Rules	Input variables		Output variables			
	VS	VW	TF	MR	TE	PC
1	VL	VL	VL	VL	VL	VL
-----	-----	-----	-----	-----	-----	-----
10	L	LM	L	L	L	L
-----	-----	-----	-----	-----	-----	-----
20	LM	H	LM	LM	LM	M
-----	-----	-----	-----	-----	-----	-----
30	HM	L	LM	HM	LM	HM
-----	-----	-----	-----	-----	-----	-----
40	H	HM	M	HM	M	VH
-----	-----	-----	-----	-----	-----	-----
49	VH	VH	VH	VH	VH	VH

Prototype triangular fuzzy sets for the fuzzy variables are set up and their membership functions and linguistic descriptions are shown in Figures 8.24-8.28. It is noted that the detection of the surface type or terrain properties for off-road machine applications is a higher-order challenge compare to automobile roads (Ivanov et al., 2010).

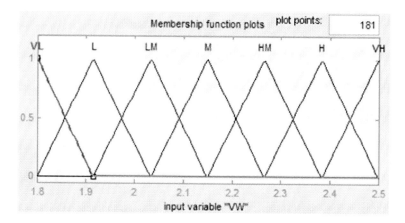

Figure 8.24. Prototype membership functions of input variable VW.

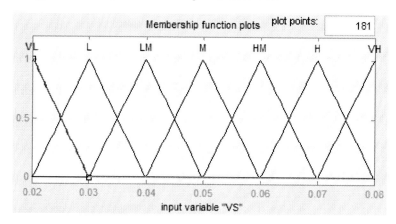

Figure 8.25. Prototype membership functions of output variable TF.

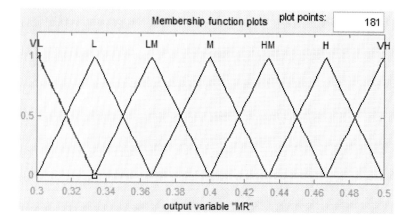

Figure 8.26. Prototype membership functions of output variable MR.

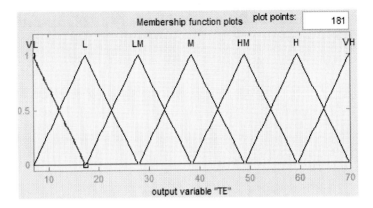

Figure 8.27. Prototype membership functions of output variable TE.

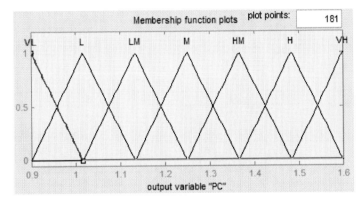

Figure 8.28. Prototype membership functions of output variable PC.

Table 8.2. Coefficients of membership functions for FIS parameter of VS

Linguistic variables	Type	Coefficients (m)		
		c_1	c_2	c_3
Very low	Z-shaped	0.02	0.03	-
Low	Triangular	0.02	0.03	0.04
Low medium	Triangular	0.03	0.04	0.05
Medium	Triangular	0.04	0.05	0.06
High medium	Triangular	0.05	0.06	0.07
High	Triangular	0.06	0.07	0.08
Very high	S-shaped	0.07	0.08	-

Table 8.3. Coefficients of membership functions for FIS parameter of VW

Linguistic variables	Type	Coefficients (kN)		
		c_1	c_2	c_3
Very low	Z-shaped	1.8	1.916	-
Low	Triangular	1.8	1.916	2.034
Low medium	Triangular	1.916	2.034	2.15
Medium	Triangular	2.034	2.15	2.266
High medium	Triangular	2.15	2.266	2.384
High	Triangular	2.266	2.384	2.5
Very high	S-shaped	2.384	2.5	-

Table 8.4. Coefficients of membership functions for FIS parameter of TF

Linguistic variables	Type	Coefficients (kN)		
		c_1	c_2	c_3
Very low	Z-shaped	0.2	0.475	-
Low	Triangular	0.2	0.475	0.75
Low medium	Triangular	0.475	0.75	1.025
Medium	Triangular	0.75	1.025	1.3
High medium	Triangular	1.025	1.3	1.575
High	Triangular	1.3	1.575	1.85
Very high	S-shaped	1.575	1.85	-

Table 8.5. Coefficients of membership functions for FIS parameter of MR

Linguistic variables	Type	Coefficients (kN)		
		c_1	c_2	c_3
Very low	Z-shaped	0.3	0.3335	-
Low	Triangular	0.3	0.3335	0.3665
Low medium	Triangular	0.3335	0.3665	0.4
Medium	Triangular	0.3665	0.4	0.4335
High medium	Triangular	0.4	0.4335	0.4667
High	Triangular	0.4335	0.4667	0.5
Very high	S-shaped	0.4667	0.5	-

Table 8.6. Coefficients of membership functions for FIS parameter of TE

Linguistic variables	Type	Coefficients (%)		
		c_1	c_2	c_3
Very low	Z-shaped	7	17.5	-
Low	Triangular	7	17.5	28
Low medium	Triangular	17.5	28	38.5
Medium	Triangular	28	38.5	49
High medium	Triangular	38.5	49	59.5
High	Triangular	49	59.5	70
Very high	S-shaped	59.5	70	-

Table 8.7. Coefficients of membership functions for FIS parameter of PC

Linguistic variables	Type	Coefficients (kW)		
		c_1	c_2	c_3
Very low	Z-shaped	0.9	1.017	-
Low	Triangular	0.9	1.017	1.134
Low medium	Triangular	1.017	1.134	1.25
Medium	Triangular	1.134	1.25	1.366
High medium	Triangular	1.25	1.366	1.484
High	Triangular	1.366	1.484	1.6
Very high	S-shaped	1.484	1.6	-

Consequently, this condition may affect on intelligent tasks for off-road and agriculture machinery. However, several options of rules bases of different sizes are studied for the input variables under consideration (Hossain et al., 2010c). The formation of membership functions is considered from the statistical data, human expertness, vehicle design parameter simulation, and so on. The type of membership functions for the input variables in the FIS may be essentially modified depending on a surface condition. The coefficients of membership functions for the input and output variables are given in Table 8.2-8.7.

To illustrate the fuzzification process, linguistic expressions and membership function of vehicle sinkage (VS) for medium (M) and high medium (HM) obtained from the developed rules with Table 8.2 is presented analytically as follows:

$$\mu_M(VS) = \begin{cases} \dfrac{x - 0.04}{0.05 - 0.04}; & 0.04 \leq x \leq 0.05 \\ \dfrac{0.06 - x}{0.06 - 0.05}; & 0.05 \leq x \leq 0.06 \\ 0; & x \rangle 0.06 \end{cases} \tag{8.21}$$

$$\mu_{HM}(VS) = \begin{cases} \dfrac{x - 0.05}{0.06 - 0.05}; & 0.05 \leq x \leq 0.06 \\ \dfrac{0.07 - x}{0.07 - 0.06}; & 0.06 \leq x \leq 0.07 \\ 0; & x \rangle 0.07 \end{cases} \tag{8.22}$$

Similarly, the linguistic expressions and membership functions of other parameters could be calculated. In defuzzification stage, truth degrees (μ) of the rules are determined for the each rule by aid of the min and then by taking max between working rules.

For example, for VS = 0.053 m and VW = 2.14 kN, the rules 24, 25, 31 and 32 are fired. To get the fuzzy inputs from FLES, VS = 0.053 m is substituted into Eq. (8.21), and obtained $\mu_M(VS) = 0.70$ and $\mu_{HM}(VS) = 0.30$ and all other membership functions are off (i.e., their values are zero). Therefore, the proposition "*VS* is *Medium*" is satisfied to a degree of 0.70, the proposition "*VS* is *High Medium*" is satisfied to a degree of 0.30; all other atomic propositions associated with *VS* are not satisfied. Similarly other two fuzzy inputs $\mu_{LM}(VW) = 0.086$ and $\mu_M(VW) = 0.914$ are obtained for VW = 2.14 kN from Figure 4.31.

The fuzzy inference system (FIS) seeks to determine which rules fire and to find out which rules are relevant to the current situation. The inference system combines the recommendations of all the rules, to come up with a final conclusion and take action. For crisp input $VS(i_1) = 0.053$ m, and $VW(i_2) = 2.14$ kN, the rules 24, 25, 31 and 32 are fired. The firing strength (truth values) of the four rules are obtained as

$$\alpha_{24} = \min\{\mu_M(VS), \mu_{LM}(VW)\} = \min(0.70, 0.086) = 0.086$$

$$\alpha_{25} = \min\{\mu_M(VS), \mu_M(VW)\} = \min(0.70, 0.914) = 0.70$$

$$\alpha_{32} = \min\{\mu_{HM}(VS), \mu_{LM}(VW)\} = \min(0.30, 0.086) = 0.086$$

$$\alpha_{33} = \min\{\mu_{HM}(VS), \mu_M(VW)\} = \min(0.30, 0.914) = 0.30$$

Therefore, the membership functions for the conclusion reached by rule (24), (25), (31) and (32) are obtained as follows:

$$\mu_{24}(TF) = \min \{0.086, \mu_{LM}(TF)\},$$

$$\mu_{24}(MR) = \min \{0.086, \mu_{M}(MR)\},$$

$$\mu_{24}(TE) = \min \{0.086, \mu_{LM}(TE)\},$$

and

$$\mu_{24}(PC) = \min \{0.086, \mu_{M}(PC)\}$$

$$\mu_{25}(TF) = \min \{0.70, \mu_{LM}(TF)\},$$

$$\mu_{25}(MR) = \min \{0.70, \mu_{M}(MR)\},$$

$$\mu_{25}(TE) = \min \{0.70, \mu_{LM}(TE)\},$$

and

$$\mu_{25}(PC) = \min \{0.70, \mu_{HM}(PC)\}$$

$$\mu_{31}(TF) = \min \{0.086, \mu_{LM}(TF)\},$$

$$\mu_{31}(MR) = \min \{0.086, \mu_{HM}(MR)\},$$

$$\mu_{31}(TE) = \min \{0.086, \mu_{LM}(TE)\},$$

and

$$\mu_{31}(PC) = \min \{0.086, \mu_{HM}(PC)\}$$

$$\mu_{32}(TF) = \min \{0.30, \mu_{LM}(TF)\},$$

$$\mu_{32}(MR) = \min \{0.30, \mu_{M}(MR)\}$$

$$\mu_{32}(TE) = \min \{0.30, \mu_{LM}(TE)\},$$

and

$$\mu_{32}(PC) = \min\{0.30, \mu_H(PC)\}$$

Defuzzification operates on the implied fuzzy sets produced by the inference mechanism and combines their effects to provide the most certain prediction output. The output membership values are multiplied by their corresponding singleton values and then are divided by the sum of membership values to compute TF^{crisp} as follows:

$$TF^{crisp} = \frac{\sum_i b_i \mu_{(i)}}{\sum_i \mu_{(i)}} \tag{8.23}$$

where b_i is the position of the singleton in the i th universe, and $\mu_{(i)}$ is equal to the firing strength of truth values of rule i. Using Equation (8.21), the crisp output of TF is obtained as 0.75 kN. Accordingly using Eq. (8.21), crisp outputs of MR, TE and PC are calculated as 0.404 kN, 28% and 1.39 kW, respectively.

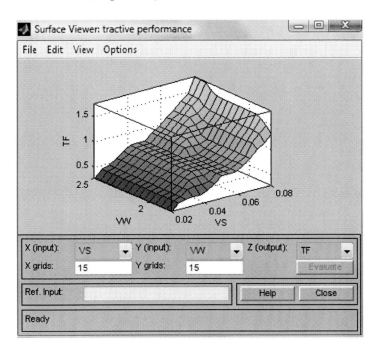

Figure 8.29. Control surfaces of the fuzzy inferring system for TF.

The fuzzy control surfaces for the set associations described in the preceding tables are shown in Figure 8.29 – 8.32, where the output variables "TF", "MR", "TE" and "PC" respectively are developed from the corresponding rules base against its two inputs. The study under consideration uses seven linguistic variables for each case to identify the TF, MR, TE and PC, respectively. The surface plots depict the impacts of the vehicle parameters on the vehicle tractive performance such as TF, MR, TE and PC, respectively. This is the mesh plot results from the interpolation of the rule base with forty nine rules. The plot is used to check the rules and the membership functions. If necessary, the rule base for the fuzzy sets is modified until the output curves are desired. Figures 8.29 – 8.32 show that the each surface represents in a compact way all the information in the fuzzy logic system. Hence, it can be noted that this representation is limited in that if there are more than two inputs it becomes difficult to visualize the surface. Furthermore, these Figures simply represent the range of possible defuzzified values for all possible inputs VS and VW. Figure 8.29 shows that as the vehicle sinkage (VS) and vehicle weight (VW) increase, there is concomitant increase in traction force (TF) and vice versa as expected.

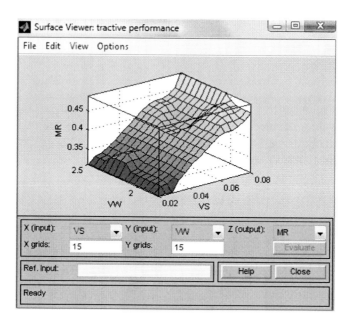

Figure 8.30. Control surfaces of the fuzzy inferring system for MR.

The traction force reaches the peak when the vehicle sinakge and vehicle weight both reach their respective maximum level, although the effect is less prominent at the higher level of vehicle sinkage due to the vehicle stuck. Consequently, traction force reaches the dip when the vehicle sinkage and vehicle weight both reach their respective minimum level.

Figure 8.30 shows that as the vehicle sinakge (VS) and vehicle weight (VW) increase, there is significant increase in motion resistance (MR) and vice versa as expected. The motion resistance reaches the peak when the vehicle sinakge and vehicle weight both reach their respective maximum level. Figure 8.30 demonstrates that as the vehicle sinkage becomes higher due to the higher load, the corresponding motion resistance increases due to the larger ground contact area.

Figure 8.31 shows that tractive efficiency reaches the peak when the vehicle sinakge and vehicle weight both reach their respective maximum level, although the effect is less prominent at the higher level of vehicle sinkage as the vehicle stuck. Consequently, tractive efficiency reaches the dip when the vehicle sinkage and vehicle weight both reach their respective minimum level. Therefore, it is important to keep the load distribution in optimum level to get the maximum tractive efficiency and hence sufficient traction to maintain the normal driving state. Decisively, it could be concluded that intelligent air cushion system plays an important role in order to reduce the vehicle normal pressure as well as increase the vehicle floatation capacity.

According to previous studies (Luo et al., 2003; Xie et al., 2009), it is observed that vehicle sinkage affects total power consumption significantly. When the vehicle with a constant load is disturbed, sinkage and required power will change. Since the total weight is supported partly by the air-cushion and partly by track system, the total power consumption is due to propulsion of tracks and for air-cushion. Hence, in order to find an optimal range of power requirements, the vehicle sinkage needs to be controlled. Figure 8.39 shows that as the vehicle sinkage increases, especially at higher level of vehicle weight, power consumption reach a peak.

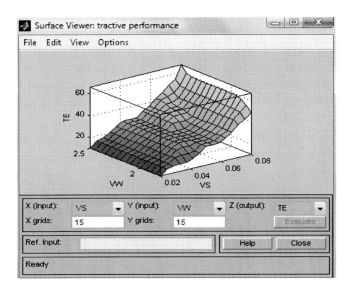

Figure 8.31. Control surfaces of the fuzzy inferring system for TE.

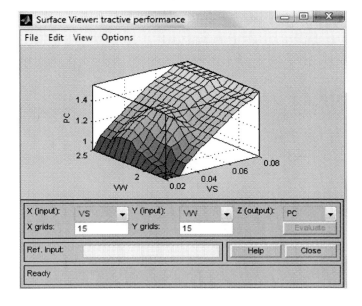

Figure 8.32. Control surfaces of the fuzzy inferring system for PC.

In contrast, the lower vehicle weight generates lower vehicle sinkage during operations and makes less dragging motion resistance, which in turn reduces the power consumption.

8.8. MODEL VALIDITY

The validation of the developed mathematical model in this study has been carried out by making comparison of the measured and predicted tractive performance of the vehicle. Prediction of tractive performance has been done by using fuzzy logic expert system (FLES) model. To validate the mathematical model, the vehicle tractive performance in terms of traction force, motion resistance, tractive efficiency and total power consumption, vehicle weight and vehicle sinkage was measured on the on soft soil (soft and wet field similar to the swamp peat) and compared with the predicted ones. For doing the proper validation of the mathematical model for the soft soil, it is important to measure the mechanical properties of the terrain and to predict the vehicle tractive performance just before the vehicle tested on the terrain. Due to the time constraint and labor intensive, it was not possible to do the field test on the terrain for determining the mechanical properties in relation to vehicle mobility. However, the terrain parameters of swamp peat were assumed to be 50% worse than that were found from the earlier research for the determination of mechanical properties in relation to vehicle mobility for Sepang peat (Rahman et al., 2004).

The effect of vehicle sinkage and weight on tractive performance such as traction force, motion resistance, tractive efficiency and total power consumption are shown in Figures 8.33-8.36. The traction force increases with increasing vehicle sinkage as well as with vehicle weight as shown in Figure 8.33. The traction force varies from 0.3925 kN to 1.4129 kN. The traction force is increasing slowly with the increase of vehicle weight until to reach a vehicle sinakge of 47 mm, and then it increases rapidly. However, the effect is found highly significant for the vehicle sinkage more than 47 mm.

This significant amount of traction force increasing is probably due to the loading situation of the vehicle and vehicle high sinkage as the cohesiveness of the field is approximately invariable for the entire terrain length.

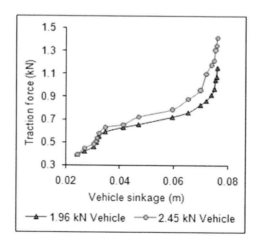

Figure 8.33. Effect of vehicle sinkage and weight on traction force.

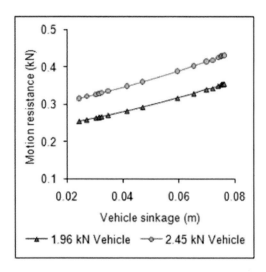

Figure 8.34. Effects of vehicle sinkage and weight on motion resistance.

Approximately, an increase of 25% at vehicle weight results in traction force increase of 18% while an increase of 61.7% at vehicle sinkage causes an 18% increase of traction force. The degree of traction force is usually larger for higher vehicle weight. Decisively, it can be concluded that the vehicle weight is the major contributory factor on traction force as compared to vehicle sinkage. The greatest value in traction force is obtained at a vehicle sinkage of 76 mm and vehicle weight of 2.45 kN.

From the motion resistance equation and previous study of Wong (2001), Luo et al. (2003) and Lyasko (2010b), it is found that vehicle sinkage and weight significantly affect on motion resistance of off-road tracked vehicles since it has direct relation with sinkage as well as vehicle weight. Figure 8.34 shows that the motion resistance increases gradually with increasing vehicle sinkage as well as with vehicle weight. The motion resistance varies from 0.2543 kN to 0.4316 kN. While the motion resistance is increasing slowly until vehicle sinkage of 35 mm, and then it increases rapidly with further increasing vehicle sinkage until to reach value of 69 mm. The gradual increasing trend is found furthermore. Similar alteration in motion resistance is found for vehicle weight as well. Approximately, an increase of 61.7% at vehicle sinkage results in motion resistance an increase of 19.62% while an increase of 25% at vehicle weight causes a 22.58% increase of motion resistance. The degree of motion resistance is larger to some extent for higher vehicle weight. However, the effect is found significant for the vehicle sinkage as well. Decisively, it can be concluded that the vehicle weight is an important contributory factor on motion resistance as compared to vehicle sinkage to some extent. The greatest value in motion resistance is obtained at a vehicle sinkage of 76 mm and vehicle weight of 2.45 kN.

A similar pattern in case of traction force is observed for tractive efficiency as shown in Figure 8.35. The tractive efficiency varies from 14.29% to 51.43%. The tractive efficiency is increasing slowly with the increase of vehicle weight until to reach a vehicle sinakge of 47 mm, and then it increases rapidly. However, the effect is found highly significant for the vehicle sinkage more than 47 mm. This significant amount of tractive efficiency increasing is probably due to the loading situation of the vehicle. Approximately, an increase of 25% at vehicle weight results in tractive efficiency increase of 22.74% while an increase of 46.81%

at vehicle sinkage causes a 21.8% increase of tractive efficiency. The degree of tractive efficiency is typically larger for higher vehicle weight. Decisively, it can be concluded that the vehicle weight is the major contributory factor on tractive efficiency as compared to vehicle sinkage (Mohiuddin et al., 2010). The greatest value in tractive efficiency is obtained at a vehicle sinkage of 76 mm and vehicle weight of 2.45 kN.

A similar pattern in case of motion resistance is observed for power consumption as shown in Figure 8.36. The power consumption varies from 0.9991 kW to 1.5648 kW. The power consumption is increasing slowly with the increase of vehicle weight until to reach a vehicle sinakge of 35 mm, and then it increases rapidly. However, the effect is found highly significant for the vehicle sinkage more than 60 mm. This significant amount of power consumption increasing is probably due to the high value of weight and hence causes greater dragging motion resistance. Approximately, an increase of 25% at vehicle weight results in power consumption increase of 22.61% while an increase of 61.7% at vehicle sinkage causes a 15.85% increase of power consumption. The degree of power consumption is naturally larger for higher vehicle weight. Decisively, it can be concluded that the vehicle weight is the major contributory factor on power consumption as compared to vehicle sinkage. The greatest value in power consumption is obtained at a vehicle sinkage of 76 mm and vehicle weight of 2.45 kN.

The results of the developed FLES have been compared with the experimental results. For traction force, the mean of measured and predicted (FLES) values have been found as 0.845 kN and 0.879 kN, respectively.

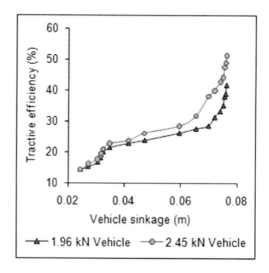

Figure 8.35. Effects of vehicle sinkage and weight on tractive efficiency.

Similarly, for motion resistance, tractive efficiency and total power consumption, the values have been found as 0.377 kN and 0.394 kN, 30.77% and 32.35%, and 1.372 kW and 1.342 kW, respectively. The correlations between measured (actual) and predicted (FLES) values of traction force, motion resistance, tractive efficiency and total power consumption in different operating conditions have been illustrated in Figures 8.37 -9.40. The relationships are significant for all the parameters in different working conditions. The correlation coefficients of traction force, motion resistance, tractive efficiency and total power consumption are found as 0.993, 0.995, 0.991, and 0.945, respectively. The mean relative error of measured and predicted values from the FLES model on traction force, motion resistance, tractive efficiency and total power consumption are found as 5.07%, 4.50%, 5.79% and 3.52%, respectively. The relative error gives the deviation between the predicted and experimental values and it is required to reach zero. For all parameters, the relative error of predicted values are found to be less than the acceptable limits of 10% (Carman, 2008). The goodness of fit of the prediction values from the FLES model on traction force, motion resistance, tractive efficiency and total power consumption are found as 0.987, 0.913, 0.981 and 0.903, respectively. The goodness of fit also gives the ability of the developed system and its highest value is 1.

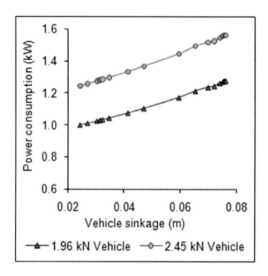

Figure 8.36. Effects of vehicle sinkage and weight on power consumption.

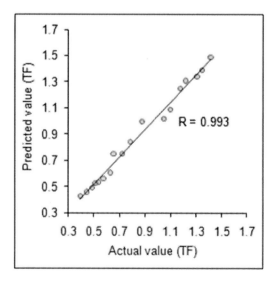

Figure 8.37. Correlation between actual and predicted values of traction force.

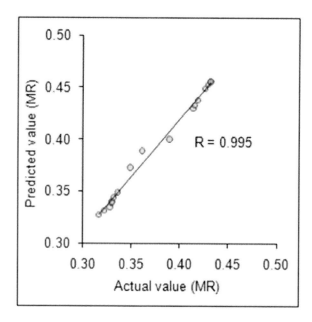

Figure 8.38. Correlation between actual and predicted values of motion resistance.

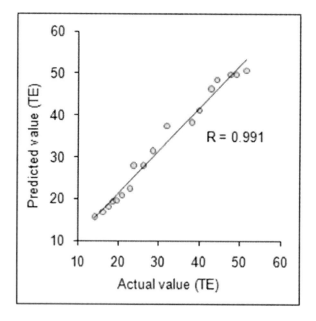

Figure 8.39. Correlation between actual and predicted values of tractive efficiency.

All values are found to be close to 1.0 as expected. However, prediction of tractive performance is necessary for agricultural engineering applications as well as automotive industries. The results indicate that there is less variability of the measured data and predicted data of the vehicle on the swamp peat terrain. It indicates that the predicted data over the measured data has a closed agreement and thus the closed agreement has substantiated the validity of the mathematical model. In this study, according to evaluation criterions of predicted tractive performance of the developed fuzzy logic expert system model has been found to be valid. Finally it can be concluded that the prediction analysis of tractive performance shows the good performance of the developed FLES model and hence warrant the novelty of this work.

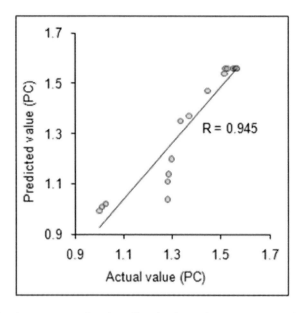

Figure 8.40. Correlation between actual and predicted values of power consumption.

CHAPTER CONCLUSION

Following Conclusion has been made for this chapter based on the intelligent vehicle:

1. For an air-cushion tracked vehicle with a given load and for a particular terrain condition, the existence of an optimal load distribution ratio which leads to a minimum power consumption, has been theoretically proved previously. Focusing on vehicle load distribution for minimizing total the power demand during normal driving, the following conclusions are made:

   - The load distribution ratio should be reasonably maintained at about 0.26 to produce sufficient traction in order to overcome the travelling resistance and maintain normal driving condition.
   - The results show that, in the present study in the case with cushion pressure p_c being 3.57 kN/m^2, slippage i within 7-25% and load distribution ratio δ being about 0.26, the minimum power consumption can be resulted.
   - When cushion clearance height is about 0.0014 m, the best overall performance of the vehicle can be obtained with relatively less cushion skirt ware.

2. Because of the auto adjusting air-cushion supporting system, load distribution can be conveniently controlled, taking vehicle vertical position h as control variable, and air

flow rate Q as regulated variable. Focusing on load distribution, the following conclusions are obtained:

- The novel auto adjusting air-cushion supporting system makes the vehicle sufficient floatation to operate on the swamp peat terrain as it allows the vehicle to traverse on any unprepared terrain by adjusting the air-cushion automatically.
- Taking vehicle position as control variable, an appropriate control scheme has been developed and its feasibility has been examined by measuring experimental data.
- By regulating volume flow rate Q through the change in valve position, the developed controller by using the fuzzy logic expert system can provide the vehicle vertical position at a desired position and hence maintains sufficient traction control.
- The mean of measured and predicted values from the FLES model on flow rate have been found as 77.78 % and 70.29 %, respectively.
- The correlation coefficient of relationship has been found as 0.971 which is significant in operation.
- The mean relative error of actual and predicted values from the FLES model on flow rate has been found as 10.93% which is almost equal to the acceptable limit of 10% (Carman, 2008).
- The goodness of fit of the prediction values from the FLES model has been found as 0.91 which is close to 1.0 as expected.

3. The intelligent air-cushion system for the vehicle has been successfully developed for traversing on swamp peat terrain as a prime mover for infield collection-transportation operation. The main special feature of the vehicle track system are: replacement segmented rubber track which could simplify servicing activities and reduce maintenance cost; longer wheelbase and shorter road-wheels interval with track system including air-cushion give better vehicle floatation, reduce vehicle sinkage, and increase vehicle traction.

4. The prediction of vehicle tractive performance in field condition is of great value for infield collection-transportation operation. Therefore, the relationships between vehicle parameters to the traction force, motion resistance, tractive efficiency and total power consumption have been investigated experimentally and illustration has been provided how fuzzy logic expert system (FLES) might play an important role in prediction of these. Focusing on vehicle tractive performance prediction, the following conclusions are drawn:

- In this study, a sophisticated intelligent model, based on Mamdani approach fuzzy logic modeling principles, has ben developed to predict the tractive performance in terms of traction force, motion resistance, tractive efficiency and power consumption.
- The mean of measured and predicted values from the FLES model have been found as 0.845 kN and 0.879 kN, respectively for traction force; 0.377 kN and 0.394 kN, respectively for motion resistance; 30.77% and 32.35%, respectively

for tractive efficiency and 1.372 kW and 1.342 kW, respectively for total power consumption.

- The correlation coefficients of traction force, motion resistance, tractive efficiency and total power consumption have been found as 0.993, 0.995, 0.991, and 0.945, respectively.
- The mean relative error of measured and predicted values from the FLES model on traction force, motion resistance, tractive efficiency and total power consumption have been found as 5.07, 4.50%, 5.79% and 3.52%, respectively.
- The goodness of fit of the prediction values from the FLES model on traction force, motion resistance, tractive efficiency and total power consumption are found as 0.987, 0.913, 0.981 and 0.903, respectively.
- The model can be used for tractive performance prediction and comparisons of off-road tracked, wheeled and air-cushion tracked vehicles in a given terrain condition without expensive and time consuming vehicle field tests.
- In general, the developed FLES model can be used for various intelligent transportation systems (ITS): identification and prediction of terrain conditions, monitoring road parameters by environmental conditions as well as prediction of track-terrain-interaction.

5. Tractive performance has been evaluated by testing the vehicle on normal surface in the laboratory and on soft terrain in the field with two different loading conditions. Focusing on vehicle tractive performance evaluation, the following conclusions are drawn:

- The mean values of the vehicle's tractive performance in terms of traction forces and motion resistances, indicate that the vehicle would not be able to traverse the swamp terrain without an air-cushion.
- The mean value of the vehicle traction forces increases with increases in the vehicle loading conditions, which could be due to the changing of the track ground contact area. Since the track width is constant, the track ground contact area is the function of the track ground contact length i.e $A_t = \mathcal{f}(L)$ where, $L = \left(L_{XY} \cos \theta + L_{YZ} + R_{rs} \sin \theta \right)$ and as well as vehicle weight.
- It is quite impossible to limit the vehicle sinkage at 70 mm if the vehicle is not equipped with an air-cushion system. Furthermore, the vehicle will get stuck in the swamp terrain if it is not equipped with an intelligent air-cushion system as the dragging motion resistance of the air-cushion system is considerable high.
- The intelligent air-cushion tracked vehicle (IACTV) has, overall, the best performance, giving about 51.6% increase in traction force as compared to air-cushion tracked vehicle without intelligent system.

REFERENCES

Al-Anbuky, A., Bataineh, S. and Al-Aqtash, S. (1995). Power demand prediction using fuzzy logic. *Control Engineering Practice*, *3*(9), 1291-1298.

Altab, H., Ataur, R., Mohiuddin, A. K. M. and Aminanda, Y. (2010). Power consumption prediction for an intelligent air-cushion track vehicle: fuzzy logic technique. *Journal of Energy and Power Engineering*, *4*(5), 10-17.

Bekker, M.G. (1956). Theory of land locomotion. Ann Arbor, MI: University of Michigan Press.

Bekker, M.G. (1969). Introduction to terrain-vehicle systems. *Ann Arbor,* MI: University of Michigan Press.

Bodin, A. (1999). Development of a tracked vehicle to study the influence of vehicle parameters on tractive performance in soft terrain. *Journal of Terramechanics*, *36*, 167-181.

Carman, K. (2008). Prediction of soil compaction under pneumatic tires a using fuzzy logic approach. *Journal of Terramechanics*, *45*, 103-108.

Castillo, O., Cazarez, N. and Rico, D. (2006). Intelligent control of dynamic systems using type-2 fuzzy logic and stability issues. *International Mathematical Forum*, *1*(28), 1371-1382.

Dwyer, M. J., Okello, J. A. and Cottrell, F. B. (1991). The effects of various design parameters on the tractive performance of rubber tracks. *Proc. ISTVS 5th Eur. Conf., Budapest.*

Dwyer, M. J., Okello, J. A. and Scarlett, A. J. (1993). A theoretical investigation of rubber tracks for agriculture. *Journal of Terramechanics*, *30*(4), 285-298.

Esch, J. H., Bashford, L.L., Bargen, K.V., and Ekstrom, R.E. (1990). Tractive performance comparison between a rubber belt track and a four-wheel drive tractor. *Transactions of the ASAE*, *33*(4), 1109-1115. ASAE, St. Joseph, MI.

Fowler, H. S. (1975). The air cushion vehicle as a load spreading transport device. *Journal of Terramechanics*, *12*(2).

George, B. and Maria, B. (1995). Fuzzy sets, fuzzy logic, applications. *Advances in Fuzzy Systems-Applications and Theory*, Vol. 5, World Scientific Publishing Co. Pte. Ltd., Singapore.

Godbole, R., Alcock, R. and Hettiaratchi, D. (1993). The prediction of tractive performance on soil surfaces. *Journal of Terramechanics*, *30*(6), 443-459.

Gopal, M. (2009). *Digital control and state variable methods: conventional and intelligent control systems.* 3rd edition, Tata McGraw-Hill Education Pvt. Ltd.

Harris, C. J., Moore, C. G. and Brown, M. (1993). Intelligent control, aspects of fuzzy logic and neural nets. London, U.K.: World Scientific.

Hossain, A., Rahman, A. and Mohiuddin, A. K. M. (2010a). Cushion pressure control system for an intelligent air-cushion track vehicle. *Journal of Mechanical Science and Technology*, (in Press).

Hossain, A., Rahman, A. and Mohiuddin, A. K. M. (2010b). Load distribution for an intelligent air-cushion track vehicle based on optimal power consumption. *International Journal of Vehicle Systems Modelling and Testing*, *5*(2/3), 237-253.

Hossain, A., Rahman, A., Mohiuddin, A. K. M. and Aminanda, Y. (2010c). Tractive performance prediction for intelligent air-cushion track vehicle: fuzzy logic approach. *World Academy of Science, Engineering and Technology*, *6*(62), 163-169.

Hossain, A., Rahman, A., Mohiuddin, A. K. M. and Aminanda, Y. (2010d). Dynamic modeling of intelligent air-cushion tracked vehicle for swamp peat. *International Journal of Aerospace and Mechanical Engineering*, *4*(1), 27-34.

Hossain, A., Rahman, A., Mohiuddin, A. K. M. and Aminanda, Y. (2010e). Development of an intelligent air-cushion tracked vehicle. *33rd FISITA World Automotive Congress 2010*, 30 May-4 June, Budapest, Hungary.

Ivanov, V., Shyrokau, B., Augsburg, K. and Algin, V. (2010). Fuzzy evaluation of tyre-surface interaction parameters. *Journal of Terramechanics*, *47*, 113-130.

Jamaluddin, B. J. (2002). Sarawak: Peat agricultural use. *Malaysian Agricultural Research and Development Institute* (MARDI), STRAPEAT, pp. 1-12, March.

Kanakakis, V., Valavanis, K. P. and Tsourveloudis, N. C. (2004). Fuzzy logic based navigation of underwater vehicles. *Journal of Intelligent and Robotic Systems*, *40*, 45-88.

Kitano, M. (1984). Development and trends in off-road vehicles in Japan. *Journal of Terramechanics*, *21*(2), 97-115.

Kitano, M. and Kuma, M. (1977). An analysis of horizontal plane motion of tracked vehicles. *Journal of Terramechanics*, *14*(4), 211-25.

Kumar, M. and Garg, D. P. (2004). Intelligent learning of fuzzy logic controllers via neural network and genetic algorithm. *Proceedings of the 2004 JUSFA, 2004 Japan-USA Symposium on Flexible Automation, Denver, Colorado*, July 19-21.

Lee, C. C. (1990). Fuzzy logic in control systems: Fuzzy logic controller-Part 1 and Part 2. *IEEE Trans. Systems, Man and Cybernetics*, *20*, 404-435.

Lee, J. (1993). On methods for improving performance of PI-type fuzzy logic controllers. *IEEE Trans. Fuzzy Systems*, *1*, 298-301.

Luo, Z. and Yu, F. (2007). Load distribution control system design for a semi-track air-cushion vehicle. *Journal of Terramechanics, 44*, 319-325.

Luo, Z., Yu, F. and Chen, B.-C. (2003). Design of a novel semi-tracked air-cushion vehicle for soft terrain. *In. J. of Vehicle Design, 31*(1), 112-123.

Lyasko, M. (2010a). LSA model for sinkage predictions. *Journal of Terramechanics, 47*, 1-19.

Lyasko, M. (2010b). Slip sinkage effect on soil-vehicle mechanics. *Journal of Terramechanics, 47*, 21-31.

Lyasko, M. (2010c). How to calculate the effect of soil conditions on tractive performance. *Journal of Terramechanics, 47*, 423-445.

Marakoglu, T. and Carman, K. (2010). Fuzzy knowledge-based model for prediction of soil loosening and draft efficiency in tillage. *Journal of Terramechanics, 47*, 173-178.

Muro, T., Brien, J. O., Kawahara, S. and Tran, D. T. (2001). An optimum design method for robotic tracked-vehicles operating over fresh concrete during straight-line motion. *Journal of Terramechanics, 38*, 99-120.

Muro, T., Tingji, H., and Miyoshi, M. (1998). Effects of a roller and a tracked vehicle on the compaction of a high lifted decomposed granite sandy soil. *Journal of Terramechanics, 35*, 265-293.

Ooi, H.S. (1993). Theoretical investigation on the tractive performance of MALTRAK on soft paddy soil. *MARDI*, Report no. 116.

Ooi, H.S. (1996). Design and development of a peat tractor prototype. *MARDI*, Report no. 184.

Park, S., Popov, A. A. and Cole, D. J. (2004). Influence of soil deformation on off-road heavy vehicle suspension vibration. *Journal of Terramechanics*, *41*, 41-68.

Park, W. Y., Chang, Y. C., Lee, S. S., Hong, J. H., Park, J. G. and Lee, K. S. (2008). Prediction of the tractive performance of a flexible tracked vehicle. *Journal of Terramechanics*, *45*(1-2), 13-23.

Passino, K. M. and Yurkovich, S. (1998). *Fuzzy control*. Addison-Wesely Longman, Inc. Menlo Park, California.

Rahman, A., Hossain, A., Mohiuddin, A. K. M., Ismail, A. F. and Yahya, A. (2010a). Integrated mechanics of hybrid electrical air-cushion tracked vehicle for swamp peat. *International Journal of Heavy Vehicle System*, *18*(1), xx-xx.

Rahman, A., Mohiuddin, A. K. M., Ismail, A. F., Yahya, A. and Hossain, A. (2010b). Development of hybrid electrical air-cushion tracked vehicle for swamp peat. *Journal of Terramechanics*, *47*(1), 45-54.

Rahman, A., Yahya, A. and Mohiuddin, A. K. M. (2007). Mobility investigation of a designed and developed segmented rubber track vehicle for Sepang peat terrain in Malaysia. *Journal of Automobile Engineering*. Proceedings of the IMech E Part D, *221*(D7), 789-800.

Rahman, A., Yahya, A., Zohadie, M., Ahmad, D. and Ishak, W. (2005). Processes involved in the design of a segmented rubber tracked vehicle for Sepang peat terrain in Malaysia. *Int. J. Vehicle Design*, *38*(4), 347-378.

Rahman, A., Yahya, A., Zohadie, M., Ahmad, D., Ishak, W. and Kheiralla, A. F. (2004). Mechanical properties in relation to vehicle mobility of Sepang peat terrain in Malaysia. *Journal of Terramechanics*, *41*, 25-40.

Rahman, Ataur., Azmi, Y., Zohadie, M., Ahmad, D. and Ishak, W. (2006a). Tractive performance of a designed and developed segmented rubber tracked vehicle on Sepang peat terrain during straight motion: theoretical analysis and experimental substantiation. *Int. J. Heavy Vehicle Systems*, *13*(4), 298–323.

Rahman, Ataur., Azmi, Y., Zohadie, M., Desa, A. and Wan, I. (2005). Design and development of a segmented rubber tracked vehicle for Sepang peat terrain in Malaysia. *Int. J. Heavy Vehicle Systems*, *12*(3), 239-267.

Rahman, Ataur., Azmi, Y., Zohadie, M., Ahmad, D. and Ishak, W. (2006b). Traction mechanics of the designed and developed segmented rubber track vehicle for Sepang peat terrain in Malaysia turning motion: theoretical and experimental analysis. *Int. J. Heavy Vehicle Systems*, *13*(4), 324-350.

Reece, A. R. (1965-66). Principles of soil-vehicle mechanics. *Proceedings of the Institution of Mechanical Engineers, 180*(2), 45-61.

Rubinstein, D. and Hitron, R. (2004). A detailed multi-body model for dynamic simulation of off-road tracked vehicles. *Journal of Terramechnics*, *41*, 163-173.

Shuib, A. and Hitam, A. (1998). Wakfoot the FFB evacuation vehicle. PORIM Information Series. *Palm Oil Research Institute of Malaysia*. PORIM TT No. 20, Pamphlet.

Shuo, X., Luo, Z., Fan, Y., Zhou, K., and Zhang, Y. (2007). Slip ratio control for a semi-track air-cushion vehicle based on power consumption optimization. *2007 IEEE International Conference on Vehicular Electronics and Safety*, Beijing, China.

Shuo, X., Zhou, K., Luo, Z., and Fan, Y. (2006). Study of fuzzy PID control for a semi-track air-cushion. *2006 IEEE International Conference on Vehicular Electronics and Safety*, Shanghai, China.

Singh, H., Bahia, H. M., and Huat, B. B. K. (2003). Varying perspective on peat, it's occurrence in Sarawak and some geotechnical properties. *Proceedings of Conference on Recent Advances in Soft Soil Engineering and Technology*, Sarawak, Malaysia, July 2-4.

Wang, D. and Qi, F. (2001). Trajectory planning for a four-wheel steering vehicle. *Proceedings of the 2001 IEEE International Conference on Robotics and Automation*, May 21-26, Korea.

Wang, X., Luo, Z., Chen, B. C., Shi, Y. W. and Ning, S. J. (1999). Study on Automatic Control System of a Semi-Tracked Air-Cushion Vehicle. *Proceedings of the IEE International Conference on Vehicle Electronics*, China.

Wik, R. M. (1984). Benjamin Holt and Caterpillar-Tracks and combines. ASAE, St. Joseph, MI. pp.129.

Wingate-Hill, R. (1975). A track-laying air cushion vehicle. *Journal of Terramechnics*, *12*(¾), 201-216.

Wong, J. Y, Garbar, M. and Preston-Thomas, J. (1984). Theoretical prediction and experimental substantiation of the ground pressure distribution and tractive performance of tracked vehicles. *Proceedings of the Institution of Mechanical Engineers*, *198*(15) D, 265-285.

Wong, J. Y. (1972a). On the applications of air cushion technology to overland transport. *High Speed Ground Transportation Journal*, *6*(3).

Wong, J. Y. (1972b). Performance of the air-cushion-surface-contacting hybrid vehicle for overland operation. *Proceedings of the Institution of Mechanical Engineers*, *186*(50/72), 613-624.

Wong, J. Y. (1998). Optimization of design parameters of rigid-link track systems using an advanced computer aided method. *Proceedings of the Institution of Mechanical Engineers*, *212*(Part D), 153-167.

Wong, J. Y. (2001). Theory of Ground Vehicles. 3rd Ed., New York, John Willey and Sons, Inc.

Wong, J. Y. (2008). Theory of Ground Vehicles. 4th Ed., New York, John Willey and Sons, Inc.

Wong, J. Y. (2009). Development of high-mobility tracked vehicles for over snow operations. *Journal of Terramechanics*, *46*, 141-155.

Wong, J. Y. and Huang, W. (2006). Wheels vs. Tracks – A fundamental evaluation from the traction perspective. *Journal of Terramechnics*, *43*, 27-42.

Wong, J. Y. and Preston-Thomas, J. (1988). Investigation into the effects of suspension characteristics and design parameters on the performance of tracked vehicles using an advanced computer simulation model. *Proceedings of the Institution of Mechanical Engineers*, *202*(D3), 143-161.

Wong, J. Y., Garber M., Radforth, J. R. and Dowell, J. T. (1979). Characterization of the mechanical properties of muskeg with special reference to vehicle mobility. *Journal of Terramechnics*, *16*(4), 163-180.

Wong, J. Y., Radforth, R. and Preston-Thomas, J. (1982). Some further studies on the mechanical properties of muskeg in relation to vehicle mobility. *Journal of Terramechnics*, *19*(2), 107-127.

Wyk, D. J. V., Spoelstra, J. and Klerk, J. H. D. (1996). Mathematical modelling of the interaction between a tracked vehicle and the terrain. *Applied Mathematical Modelling, 20*, 838-846.

Xie, D., Luo, Z. and Yu, F. (2009). The computing of the optimal power consumption for semi-track air-cushion vehicle using hybrid generalized external optimization. *Applied Mathematical Modelling, 33*, 2831-2844.

Xie, D., Ma, C., Luo, Z. and Yu, F. (2008). Pitch control for a semi-track air-cushion vehicle based on optimal power consumption. *World Automotive Congress, (FISITA 2008)*, Germany, F2008-01-009, 14-19 September.

Zadeh, L. A. (1965). Fuzzy sets. *Information and Control, 8*, 338-353.

Zadeh, L. A. (1973). Outline of a new approach to the analysis of complex systems and decision processes. *IEEE Trans. Systems, Man and Cybernetics, 3*, 28-44.

Zoz, F. M. (1970). Predicting tractor field performance. *ASAE* Paper No. 70-118, ASAE, St. Joseph, MI 49085, 1-11.

Chapter 9

SEMI-WHEELED TRACKED VEHICLE: "HIGHLAND AND MODERATE PEAT"

9.1. INTRODUCTION

Typical peat characteristics as found in Malaysia is very low bearing capacity terrain and it is composed of submerged and un-decomposed woods, stumps and logs. These submerged and unrecompensed logs or stumps impede the movement of machinery in the field. Other important characteristics are the very high ground water table, low bulk density and bearing capacity. The Sepang peat terrain bearing capacity of 12 kN/m^2 is considered for studying the traction mechanics of the proposed vehicle Ref.[1]. Under the loaded surface some of peat soil may be at rest while others may move down. It is very difficult to manage any vehicle such as wheeled vehicle and tracked vehicle. It is noted due to the much ground contact pressure of wheel it is absolutely to possible to operate. While, the tracked vehicle is good for strain running but difficult to maneuver over the terrain due to its high lateral sliding which bogged down the vehicle. Furthermore, the vehicle's mobility is limited by the terrain capacity and climatic affect. In general, if the vehicle ground contact pressure is more than the normal ground pressure of the terrain, the vehicle is at risk to operate on the low bearing capacity peat terrain. Therefore, this study is conducted to solve the problems and increase the vehicle overall performance by decreasing the track ground contact length and the tire inflation pressure to ease the vehicle maneuverability. The main advantage of semi-tracks wheel vehicle over the wheeled vehicles is that the tracks reduce the vehicle's overall ground contact pressure and give it greater mobility over soft and steep terrain. While, this vehicle does not require the complex steering mechanisms of fully tracked vehicles, relying instead on its front wheels to direct the vehicle, augmented in some cases by track braking controlled by the steering wheel. This is a hybrid electromagnetic engine powered vehicle. The small conventional engine is considered for this vehicle. The engine rpm keeps constant in order to produce the desired power. A 24 volts and 67 N.m @ 300 rpm DC motor will be used with this vehicle tracked system to develop torque and speed immediately for developing the vehicle's traction in any terrain condition. The main objective of use the hybrid engine to make the vehicle as a green vehicle with reducing the vehicle dry load and emission like CO_2 and N_2O. Therefore, the vehicle could be considered as a green vehicle.

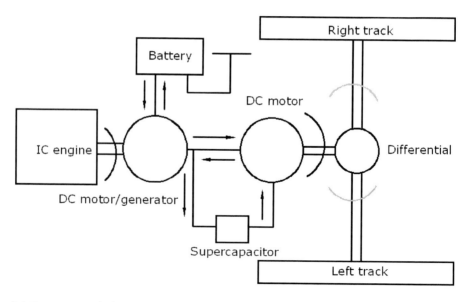

Figure 9.1. Power transmission system.

9.2. Traction Mechanics

The kinematics model of this vehicle is developed with comprehensive understanding of the traction mechanics of wheeled and tracked system individually. The mathematical model is then used for simulating the vehicle tractive performance, optimizing the design parameters and load distribution from the vehicle's track system to the wheel. The power train of the vehicle as shown in Figure 1 is developed in such a way that its power loss due to the braking will be recovered with regenerative braking system. The super capacitor is considered as the major power source which is charged by the power of hybrid engine and its regenerative braking system and discharged during operation of the vehicle. The Hybrid engine of this vehicle is able to produce much amount of power as required of the vehicle propel in any adverse condition.

Based on the study of traction of the wheeled and tracked system, it is found that the tracked system for the tracked vehicle and LPG-30 wheeled for the wheel vehicle is highly potential to operate on the low bearing capacity terrain of bearing capacity 12 kN/m^2. Therefore, tracked of width 0.335 m and 0.06 m grouser height for the tracked system and LPG-30 wheeled vehicle are considered for the proposed vehicle tracked system and wheel. The main purpose of consider the oval tracked system just to increase the proposed vehicle ground clearance. The centre line of the wheeled and the driving sprocket will be placed in the same line as shown in Figure 1 to maintain the turning maneuverability of the vehicle in any terrain situation and to increase the mobility of the tracked system. It could be stated that the wheeled of the vehicle will compress the terrain with minimum sinkage as the load on the wheel system is much lower than the tracked system.

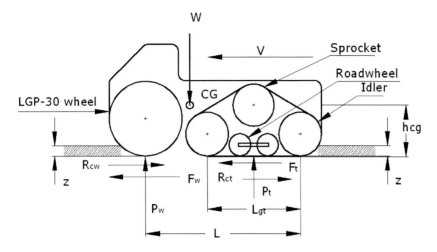

Figure 9.2. Kinematics of semi-tracked wheeled vehicle.

The vehicle tractive performance is analysed by taking accounts the vehicle's weight, weight distribution, wheel and track contact length, sinkage, slippage, and centre of gravity (CG) location.The development of the mathematical model in this study is initiated by classifying into: (i) kinematic model for the wheeled and tracked system based on the slippage, (ii) the algorithm modeling for the tyre-terrain and track-terrain interaction by simplifying the general tractive and motion resistance equations reported in ref.[1-6]. In this study, peat terrain is considered as the low bearing capacity terrain.

9.2.1. Traction Force

The mathematical model for the semi-tracked wheeled vehicle is developed with integrated the mathematical model of the tracked system and wheeled system. The tractive effort of the semi-wheeled tracked vehicle is computed by using the equation of *Ataur et.al* (2005) and (2009).

$$F_t = (c + \sigma_t \tan \varphi) \left[\begin{array}{l} (4H_t L_t \cos \alpha + 2B_t L_t)(multiplying\ factor)_t \\ + \left(\dfrac{\sigma_w B_w L_w}{\sigma_t \sin \varphi_1} \right)(multiplying\ factor)_w \end{array} \right] \quad (9.1)$$

where

$$\sigma_t = \dfrac{W_t}{B_t L_t}$$

and

$$\sigma_t = \frac{W_w}{B_w L_w}$$

$$(multiplyin\ g\ factor)_t = \left[\frac{K_w}{i_t L_t}e^1 - \left(1 + \frac{K_w}{i_t L_t}\right)\exp\left(1 - \frac{i_t L_t}{K_w}\right)\right]$$

$$(multiplyin\ g\ factor)_w = \left[\frac{K_w}{i_w L_w}e^1 - \left(1 + \frac{K_w}{i_w L_w}\right)\exp\left(1 - \frac{i_w L_w}{K_w}\right)\right]$$

In equation (8.1), F_t is the tractive effort of the vehicle in kN, c is the cohesiveness of the terrain in kN/m^2, σ_t and σ_w are the normal stress of the vehicle on the track-terrain and wheel-terrain interaction in kN/m^2, L_t and L_w are the ground contact part of the track and wheel in m, B_t and B_w are the track and wheel width in m, i_t and i_w the slippage in percentage, K_w the shear deformation modulus in m, i is the slippage of the track in percentage, H is the height of the grouser in m, and α is the angle of the track system between the grouser and track width in degree.

9.2.2. Motion Resistance

The motion resistance of the proposed vehicle is computed by considering the negligible effect of aerodynamic motion resistance as the vehicle will travel at speed of 12 km/h and belly drag as the vehicle ground clearance. The motion resistance which is mainly incurred due to the terrain compaction can be shown as,

$$R_c = 2B\left(\frac{\left(z_{0t}^2 + z_{0w}^2\right)k_p}{2} + \frac{4}{3\left(D_{ht} + D_{hw}\right)D_h}m_m\left(z_{0t}^3 + z_{0w}^3\right)\right)(9.2)$$

where

$$D_{ht} = \frac{B_t H_g}{\left(H_g + B_t\right)}\ and\ D_{hw} = \frac{B_w H_c}{\left(H_c + B_w\right)}$$

In equation (9.2), R_c is motion resistance due to terrain compaction in kN, z_{ot} and z_{ow} are the sinkage of the tracked and wheeled of the vehicle in m, B_t and B_w are the width of the track and wheel in m, H_g is the grouser height of the track and H_c is the carcass height of the wheel in m, D_{ht} and D_{hw} are the hydraulic diameter of the track and wheel respectively. Table 9.1 shows the design parameters of the semi-tracked wheeled vehicle which are considered based on the traction of the LGP wheel and track system.

Table 9.1. Proposed vehicle design parameters

Vehicle parameters		
Weight, kN	W	24.5
Wheel base, m	L	2.45
Ground clearance, m	G_c	0.48
Load carrying capacity, m	W_p	
Ground contact length, m	L_t	1.50
Track width,m	B	
Vehicle distributed weight on tracked system, kN	W_t	16.82
Wheeled system		
Tyre 16x24 8PRTube Type		
Wheel diameter, m	D_w	1.21
Vehicle distributed weight , kN	W_w	7.71

Chapter 10

RELEVANT PEAT VEHICLE IN MALAYSIA

10.1. INTRODUCTION

There are some vehicle has been locally for the mechanisation of agriculture on peat terrain. Figure 10.1 shows a Prototype Track Type Tractor is a machine designed by Ooi (1996) and developed by local agricultural machinery manufacturer for using in cassava mechanization on peat as shown in Figure 9.2. This fully hydrostatic machine is fitted with front-wheel sprocket driven chain link wooden grouser tracks, which give low ground pressure and high floatation for good performance in peat terrain.

Source : MARDI,1996 Report No. 184.

Figure 10.1. Peat tractor prototype (Source: MARDI, 1996 Report No. 184).

It is powered by a two cylinder diesel engine of 41.79 kW coupled to a hydrostatic pump that operates two hydrostatic motors through two gear boxes, one for each sprocket that drives the metal track. The tension of these tracks is self-adjusted by means of springs. Together with the 9.81kN weight of the vehicle, the ground pressure exerted is about 17.48kN/m^2. Steering of this machine is by means of two levers to control the movement forward, backward, 'stop', left or right. Table 10.1 shows the technical specifications of the Peat Prototype Track Type Tractor. The position of PTO and three-point linkage assembly was fixed directly behind and below of the tractor operator, near enough to give him a clear view and good control of the implements and attachments used.

Table 10.1. Technical specifications of peat tractor prototype

Maximum loading capacity, kN	17.65
ENGINE	
Model	Lomberdini
Type	Air cooled
No. of cylinders	4
Power,kW@ RPM	45 @ 2000
VEHICLE DIMENSIONS	
Length, mm	3000
Width, mm	1560
Height, mm	1250
Min. ground clearance, cm	40
TRACK PARAMETERS	
Ground contact length, mm	1650
No. of lower track rollers	3
No. of upper track rollers	2
Center-to-center distance,mm	1100
Track width, mm	460
OPERATING PARAMETERS	
Ground pressure fully loaded,kN/m^2	17.46
Max. Traveling Speed,km /hr	10
No. of Speeds, forward/reverse	3 / 3

10.2. PEAT TRACTOR PROTOTYPE

Field performance evaluation of the peat tractor prototype in rotary slashing and rotary tillage showed that the machine was suitable for field clearing work and for soil preparation work under field conditions existing at MARDI peat research station at Jalan Kebun.

In order to avoid the friction driven the peat tractor prototype was designed to be front wheel sprocket driven for getting direct positive drive instead of being rear wheel driven. The track system of this machine transmitted more vibration and made discomfort to the operator than rubber tracks. The problem of hydraulic oil overheating after 2 hours of continuous operation which will damage the oil seal of the hydraulic system and increases the leakage of the oil.

Figure 10.2. Photo of MALTRAK (Source:MARDI, 1993, Report no. 166).

10.2. MALTRAK

The MALTRAK was designed and developed by Ooi (1993) for light rotary tillage, spraying, broadcasting and transporting 9.81kN load on developed peat terrain as shown in Figure 10.2. This machine is fitted with 450mm continuous rubber belt that wraps round three 690mm diameter light truck wheels for low ground pressure and high floatation on peat terrain. The track belt is metal cleated at 150mm spacing with wheel retainer cum grip plates on the inside and 40 mm high grousers on the outside.

Table 10.2. Technical specifications of MALTRAK

MACHINE WEIGHT	16.67
Machine weight empty ,kN	12.75
Maximum loading capacity, kN	
ENGINE	
Model	Lombardini
Type	Air cooled
No. of cylinders	4
Power,kW@ RPM	32 @ 2000
Vehicle dimension	
Length,mm	2890
Width, mm	1100
Height,mm	1740
Min. ground clearance,mm	400
TRACK PARAMETERS	
Ground contact length, mm	1700
No. of pneumatic tire wheels	3
Pneumatic wheels diameter,mm	690
Center-to-center distance,mm	1100
Track width,mm	450
OPERATING PARAMETERS	
Ground pressure fully loaded,kN/m^2	19.62
Max. Traveling Speed,km /hr	10
No. of Speeds, forward/reverse	3 / 3

REFERENCES

Okello, J.A., Watany, M., and Crolla, D.A. 1998. Theoretical and Experimental Investigation of Rubber Track Performance Models, *Journal of Agricultural Engineering Res.*Vol. 69, pp.15-24.

Ooi, H.S. 1993. Theoritical investigation on the tractive performance of MALTRAK on soft padi soil. MARDI, Report no.116.

Ooi, H.S. 1996. Design and development of peat prototype track type tractor. MARDI, Report no.184.

INDEX

A

access, 8, 62
accessibility, 28, 29
accounting, 6, 25, 120
acid, 9, 12, 38, 139
acidic, 3, 4, 6
acidic water regimes, 3
acidity, 22, 41
adaptation, 8
adjustment, 90
aesthetic, 28, 29
agriculture, 6, 8, 12, 15, 17, 22, 23, 24, 28, 29, 38, 65, 66, 117, 118, 150, 164, 175
Air Hitam Laut village, 16
Alaska, 25
algorithm, 165, 171
ANOVA, 46, 48, 55, 60
appraisals, 147
aquarium, 23, 25, 26
aquifers, 5
artificial intelligence, 145
ASEAN, 7, 8, 20, 21, 22, 37, 38, 61
Asia, 12, 13, 14, 23, 30, 61
Asian countries, 10
assessment, 8, 14, 31, 41
assets, 16, 19
atmosphere, 4, 9, 10, 11, 26
automate, 118
Automobile, 87, 88, 115, 166
awareness, 29

B

bacteria, 26
barriers, 12
base, 21, 39, 40, 123, 130, 145, 147, 153, 173
batteries, 138, 139

Beijing, 166
Belarus, 27, 28, 30
benefits, 17, 19, 21
bias, 77
biodiversity, 21, 22, 28
biological processes, 7, 20
biomass, 10, 11
breeding, 26
building blocks, 16
burn, 11, 15

C

canals, 11, 23, 38
capillary, 13
carbon, 5, 9, 10, 11, 14, 19, 21, 22, 26, 27, 30, 31, 37, 38, 39, 40
carbon dioxide, 37
case study, 14, 21, 30
cash, 27, 28
catchments, 5
cation, 9, 39, 40
cattle, 29
cellulose, 40
Central Asia, 27
challenges, 15, 16
chemical, 7, 17, 20, 38, 39
chemical characteristics, 38
chicken, 25
chicken pox, 25
China, 166, 167
circulation, 4
cities, 24
classification, 41
climate, 38
CO_2, 9, 10, 11, 12, 13, 26, 27, 30, 61, 169
coal, 9
collaboration, 16

combustible peat matter, 3
commercial, 19, 22, 23, 24, 25, 118
commercial crop, 23
community(s), 15, 16, 17, 18, 23, 24, 25, 26, 28, 29
compaction, 22, 76, 82, 93, 96, 128, 129, 137, 146, 164, 165, 172
complexity, 90
composition, 3
compost, 40
composting, 14, 40, 62
compressibility, 35, 42, 61
compression, 42
computation, 74
computer, 114, 167
computing, 74, 91, 112, 168
conditioning, 40
conductivity, 13, 40
configuration, 90, 110
Congress, 165, 168
conservation, 12, 31, 38
conserving, 31
consolidation, 40, 47, 52, 54, 57, 60
constant load, 134, 155
construction, 8, 17, 27, 35, 77, 117, 118, 145
consumption, 118, 130, 134, 135, 137, 145, 146, 147, 155, 156, 158, 159, 161, 162, 163, 164, 166, 168
cooking, 16
correlation, 146, 158, 162, 163
correlation coefficient, 146, 158, 162, 163
correlations, 158
cosmetics, 17
cost, 36, 37, 65, 89, 90, 110, 162
covering, 3, 52
crop, 6, 10, 11, 17, 19, 23
Crop Research and Application Unit (CRAUN), 17
crops, 6, 11, 13, 17, 18, 23, 24, 39
cultivation, 6, 10, 17, 18, 24, 37, 39
culture, 28
currency, 27
cycling, 3

D

danger, 15
data set, 145
decay, 6
decomposition, 3, 9, 11, 12, 27, 39
deforestation, 10
deformation, 35, 41, 42, 52, 54, 56, 70, 71, 75, 76, 147, 166
degradation, 10, 15, 17, 26
deposits, 26, 31, 35, 40

depth, 4, 9, 10, 11, 12, 37, 38, 41, 42, 46, 47, 48, 50, 51, 54, 59, 68, 72, 128, 146
designers, 65
destruction, 9
detection, 148
developing economies, 29
deviation, 131, 159
direct observation, 41
discharges, 28
discomfort, 176
displacement, 52, 53, 70, 71, 75, 96, 100, 101, 119, 125, 127
dissolved oxygen, 3
distribution, 25, 72, 74, 80, 87, 89, 90, 91, 94, 98, 115, 117, 120, 121, 122, 124, 127, 133, 134, 135, 137, 154, 161, 164, 165, 167, 170, 171
diversity, 25, 36, 65
draft, 165
drainage, 5, 8, 10, 11, 12, 13, 22, 23, 24, 27, 29, 37, 38, 44, 46, 47, 48, 51, 52, 54, 55, 58, 59, 143
drawing, 142
drought, 14
drying, 3, 41
dyeing, 27
dynamic systems, 164

E

East Asia, 61
Eastern Europe, 27
ecology, 29
economic development, 38
economic losses, 29
economic systems, 6, 7, 20
ecosystem, 3, 5, 6, 7, 15, 17, 19, 20, 29, 38, 62
education, 29
electromagnetic, 83, 169
emission, 9, 10, 11, 12, 13, 26, 61, 169
employment, 27
endangered, 20, 25
endangered species, 20, 26
energy, 3, 27
engineering, vii, 8, 9, 41, 132, 160
environment, 3, 8, 19, 107, 119, 132
environmental change, 19, 20
environmental conditions, 163
environmental degradation, 19
environmental effects, vii
environmental impact, 19
environmental resources, 19
environmental services, 15
environments, 5, 15, 132
equilibrium, 72

Index 181

equipment, 25
erosion, 17
Eurasia, 27
Europe, 9, 23
European Commission, 30
evacuation, 12, 166
evaporation, 13
Everglades, 28, 29
evidence, 30
evolution, 3
exploitation, 8, 16
exporter(s), 24
external environment, 15
extraction, 5, 21, 25, 29

F

farmers, 37, 38
farms, 6, 17
fauna, 5, 18, 31
fermentation, 14, 39, 40, 62
fertilization, 22
field tests, 41, 163
financial, 16, 19
financial resources, 16
Finland, 27, 28, 29
fires, 10, 16, 24, 29, 37
fish, 4, 5, 23, 25, 26, 31
fisheries, 3, 17, 38
fishing, 17, 18, 25
flatness, 6
floatation capabilities, 36, 65
flood mitigation, 3
flooding, 3, 7, 12, 20, 28
floods, 5, 15, 28
flora, 3, 5, 30
folklore, 28
food, 10, 12, 17, 21, 22, 24, 27, 39
food production, 12
force, 43, 67, 68, 70, 71, 72, 74, 75, 76, 80, 82, 83,
  84, 85, 86, 95, 96, 97, 124, 125, 126, 127, 128,
  133, 134, 135, 137, 138, 140, 141, 142, 143, 144,
  145, 146, 147, 154, 156, 157, 158, 159, 162, 163
forest ecosystem, 38
forest management, 21
formation, 3, 5, 150
formula, 147
freshwater, 5
friction, 41, 52, 53, 54, 56, 60, 70, 77, 85, 87, 92, 95,
  96, 125, 127, 128, 130, 137, 176
fruits, 27, 117
fuel consumption, 26
fungi, 3

fungus, 14, 40, 62
fuzzy set theory, 145
fuzzy sets, 145, 148, 153, 154

G

geometry, 72
Germany, 23, 28, 29, 61, 90, 168
GHG, 8, 9, 10
Global Environment Facility, 61
global scale, 5
global warming, 5, 9
goods and services, 3, 19, 21
GPS, 98
graph, 46, 54
grass(s), 3, 4, 35, 38, 43, 142, 143
gravity, 12, 52, 90, 91, 97, 120, 122, 147, 171
greenhouse, 9, 10, 11, 14, 26
greenhouse gas(s), 9, 10, 11, 14, 26
greenhouse gas emissions, 14
groundwater, 3, 5, 11, 13, 22, 28
growth, 6, 9, 17, 40
growth rate, 9

H

habitat(s), 5, 20, 22, 25, 28, 29
harvesting, 19, 21, 25
hay meadows, 27
hcg, 91, 122
health, 19
height, 11, 58, 77, 91, 93, 110, 117, 120, 122, 127,
  128, 131, 132, 134, 145, 147, 161, 170, 172
herbicide, 40
highlands, 35
history, 29
HM, 147, 151
House, 8
housing, 35
human, 5, 15, 21, 25, 29, 117, 150
Hungary, 165
hybrid, 166, 167, 168, 169, 170

I

identification, 163
illegal logging, 15, 16
impact assessment, 8
imports, 27
incidence, 11
income, 16, 17, 25, 27, 28, 29
income generating activities, 16

independence, 27
individuals, 15, 19, 20, 22
Indonesia, 9, 10, 14, 18, 23, 24, 25, 26, 28, 29, 31, 35, 39, 62
industry(s), 17, 23, 25, 35, 39, 40, 117, 145, 160
inflation, 67, 68, 71, 83, 84, 85, 86, 87, 109, 131, 145, 169
infrastructure, 7, 20, 29, 35, 38
insects, 3
institutions, 16
integration, 147
interaction process, 146
interface, 68, 70, 71, 91, 124
interrelations, 36, 65
Ireland, 27
irrigation, 8, 20, 22, 39
isolation, 16
issues, 164

J

Japan, 165
justification, 120

K

kerosene, 52
Korea, 167
Kyoto Protocol, 30

L

Land Custody Development Authority (LCDA), 17
landscape, 3, 28
laptop, 98
lateral motion, 96
lead, 10, 12, 139
leakage, 176
learning, 165
light, 177
local community, 15
local conditions, 16
locus, 69
logging, 3, 15, 16, 17, 21, 38
low shrubs, 43
LPG, 170
LSD, 47, 48, 49, 54, 59, 106
Luo, 118, 120, 121, 133, 135, 145, 155, 157, 165, 166, 167, 168
lying, 6, 17, 35

M

machine learning, 145
machinery, 8, 67, 150, 169, 175
majority, 13, 38
Malaysia, v, 6, 7, 9, 10, 14, 17, 18, 21, 23, 24, 25, 26, 28, 29, 30, 31, 35, 36, 41, 42, 62, 67, 83, 88, 89, 90, 114, 115, 118, 141, 166, 167, 169, 175
mammals, 25
man, 19
management, 9, 10, 12, 14, 21, 22, 30, 31, 132
mangroves, 3
manipulation, 8
manufacturing, 8, 90
mapping, 145
mass, 28, 29, 49
materials, 12, 17, 35, 40, 47, 54, 90, 119, 132
matrix, 29
matter, 3, 38, 39
measurement, 11, 41, 49
meat, 24
mechanical properties, vii, 41, 60, 61, 62, 65, 67, 77, 78, 109, 112, 114, 156, 167
medicine, 21, 24, 27
membership, 147, 148, 149, 150, 151, 152, 153, 154
memory, 147
metabolic pathways, 40
metabolized, 40
meter, 41, 72, 83, 98, 101
methodology, 119, 120
microclimate, 5
migrants, 11
migration, 16
military, 117, 118
mission, 8
modelling, 168
models, 65, 109, 110, 111, 112, 119, 120, 132, 145
modifications, 78
modulus, 41, 52, 53, 54, 56, 60, 77, 85, 92, 95, 96, 125, 127, 128, 172
moisture, 3, 9, 11, 12, 41, 49, 50, 51, 60, 61, 77, 87, 105, 106, 107
moisture content, 9, 11, 12, 41, 49, 50, 51, 60, 77, 87, 105, 106, 107
MR, 147, 148, 150, 153, 154

N

natural resources, 16, 17
natural sponge, 3
nature conservation, 18
negative effects, 16

net present values, 21
Netherlands, 16, 28, 29, 114
neural network, 145, 165
nitrogen, 10, 39
nitrogen fixation, 39
nitrous oxide, 10
North America, 24
nutrient(s), 3, 6, 9, 17, 22, 24, 38, 39

O

obstacles, 16, 121
oil, 6, 7, 10, 11, 13, 16, 17, 23, 24, 27, 30, 36, 39, 40, 61, 112, 117, 176
oil production, 30
operations, 8, 9, 89, 117, 134, 156, 167
opportunities, 6, 15, 16, 17, 23
optimization, 8, 109, 110, 112, 166, 168
organic compounds, 39
organic matter, 3, 6, 38, 39
organic soils, 35, 41, 42
ornamental plants, 25
oxidation, 9, 10, 12, 13, 27
oxygen, 4, 37, 38

P

palm oil, 10, 14, 24, 40, 62, 66, 67, 89
palm roots, 43
pasture, 37
pastures, 24, 27
peatland ecosystem, 7, 15, 17, 20, 29, 37
pH, 4, 39, 40, 41
pharmaceutical, 17
phosphorus, 39
physical characteristics, 39, 147
pitch, 91, 110, 120, 122
plants, 25, 27, 30, 39
Poland, 27
policy, 15
policy levels, 15
pollen, 29
pollutants, 28
pollution, 9, 19, 29
population, 10, 12, 16
porosity, 40
poverty, 15, 16, 23, 24
poverty reduction, 15
precipitation, 11
preparation, 176
preservation, 38
principles, vii, 30, 118, 162

probability, 46, 48, 55, 60, 106, 108
project, 9, 15, 16, 30
proposition, 151
protection, 8, 19, 117
prototype(s), 65, 115, 118, 135, 166, 175, 176, 178
pumps, 99
purification, 21

R

radar, 83
radius, 58, 65, 69, 70, 71, 91, 93, 94, 110, 113, 122, 124, 127, 128
rain forest, 20
rainfall, 3, 5, 12, 13, 38
reality, 119, 132
recommendations, 111, 151
recreation, 18, 19, 22, 28, 29, 30
recreational, 21
regeneration, 25
rehabilitation, 28, 29
reliability, 36, 65
requirements, 12, 36, 65, 89, 155
research institutions, 16
reserves, 28, 29
resolution, 57
resource allocation, 20
resources, 5, 17, 19, 20, 21, 24, 29
respiration, 9, 12
response, 5, 65
revenue, 25
rhino, 25
risk(s), 9, 12, 16, 17, 20, 38, 49, 68, 78, 90, 117, 139, 169
rods, 41, 57
root(s), 3, 4, 5, 9, 11, 12, 29, 35, 43, 112, 117
rotational inertia, 67
rubber, 10, 11, 36, 88, 89, 91, 109, 110, 114, 115, 118, 139, 162, 164, 166, 176, 177
rules, 119, 130, 131, 132, 145, 147, 150, 151, 153
rural areas, 27
rural population, 29
Russia, 27, 35

S

safe haven(s), 25
safety, 28
saline water, 8, 20
saturation, 39, 40, 43
science, 26, 114
security, 29

sediment(s), 3, 6, 9
sensors, 98
services, 15, 16, 19, 21, 29, 38
shape, 12, 112
shear, 35, 40, 41, 52, 53, 54, 56, 57, 58, 59, 60, 77, 85, 92, 95, 96, 101, 110, 124, 125, 127, 128, 147, 172
shear deformation, 41, 52, 53, 54, 56, 60, 77, 85, 92, 95, 96, 125, 127, 128, 172
shear strength, 35, 41, 53, 54, 58, 110
showing, 78
shrubs, 43, 119
signs, 15
simulation, 43, 77, 78, 82, 87, 112, 135, 136, 150, 166, 167
Singapore, 30, 164
social costs, 19
society, 19
solid state, 14, 40, 62
South Africa, 24
Southeast Asia, 14, 24, 25, 26, 31, 35, 62
species, 3, 4, 5, 23, 24, 25, 27, 30, 31
specific gravity, 52
specifications, 77, 175, 176, 177
spin, 79
sponge, 3, 5, 7, 20
SS, 46, 48, 55, 60
stability, 97, 110, 113, 164
stakeholders, 29
state(s), 6, 7, 12, 19, 22, 23, 25, 27, 35, 65, 68, 110, 154, 164
steel, 112
stock, 10, 14
storage, 5, 9, 12, 13, 21, 22, 26
stress, 53, 54, 55, 72, 75, 92, 95, 96, 125, 127, 172
structure, 35, 40, 41
subsistence, 6, 17, 24, 28
substrate, 5, 40
sulfuric acid, 9
surface mat, 35, 41, 43, 45, 46, 47, 60, 68, 75, 77, 78, 87, 90, 94, 110, 113, 118, 124
surplus, 5
survival, 3
sustainability, 22
sustainable development, 14
Sustainable Development, 14, 31
Sweden, 27, 114
Switzerland, 31
system analysis, 36, 65

T

target, 16, 23, 36

target population, 16
techniques, 16, 35, 60, 145
technology, 118, 167
temperature, 4, 13, 49, 52
tension(s), 29, 47, 90, 113, 120, 175
test procedure, 44, 52, 58
testing, 60, 61, 83, 111, 118, 141, 143, 146, 163
Thailand, 25, 62
threats, 37
tin, 23
Tonga, 18
torsion, 90
tracks, 36, 65, 89, 90, 94, 95, 96, 101, 109, 114, 117, 118, 128, 129, 139, 140, 141, 155, 164, 169, 175, 176
traditions, 28
training, 130
transducer, 83
transmission, 36, 65, 77, 97, 139, 140, 170
transport, 38, 89, 164, 167
transportation, 8, 18, 35, 36, 38, 66, 67, 89, 110, 118, 119, 162, 163
treatment, 6, 8, 13, 40, 41, 61
tropical forests, 26
tropical monsoon, 3

U

UK, 28, 29, 30, 31, 88, 115
underwater vehicles, 165
UNDP, 9
uniform, 90, 91, 117, 127, 135
universe, 153
universities, 118
USA, 28, 29, 165

V

vacuum, 52
validation, 86, 156
valuation, 19
valve, 130, 131, 132, 145, 146, 162
variables, 130, 145, 147, 148, 149, 150, 153
variations, 50, 59, 107, 135
varieties, 5
vegetables, 24, 27
vegetation, 3, 5, 6, 9, 11, 18, 24, 25, 26, 38, 90
vehicles, 8, 65, 66, 80, 87, 89, 90, 106, 109, 114, 115, 117, 118, 157, 163, 165, 166, 167, 169
velocity, 6, 28, 67, 83, 97, 98, 106, 107, 125, 130, 135, 145
vibration, 113, 166, 176

Index

Vietnam, 9, 11, 49, 61
vulnerability, 15

W

walking, 43
waste, 40
water, 3, 4, 5, 6, 7, 8, 9, 11, 12, 13, 14, 16, 17, 20, 21, 22, 24, 25, 26, 28, 30, 35, 37, 38, 39, 41, 42, 43, 47, 49, 52, 54, 60, 67, 106, 107, 112, 118, 143, 169
water purification, 5
water quality, 30, 37
water supplies, 22
waterlogged environment, 3
watershed, 17, 18, 19
web, 38
weight ratio, 36, 65, 110
well-being, 19, 20, 27
wetlands, 3, 11, 26, 27, 30, 31, 37
wilderness, 19
wildlife, 21
wildlife conservation, 21
wood, 25, 27, 134
wood products, 25
working conditions, 89, 112, 146, 158
World Bank, 30
World War I, 90
worldwide, 5, 27

Y

yield, 27